The Instrumental Spectrometric and Spectroscopic Analysis of Amino Acids

—

A Handbook

$$R - \underset{\underset{NH_2}{|}}{\overset{\overset{H}{|}}{C}} - COOH$$

Jasmine Tripconey

and

Nick Winstone-Cooper

First published by Ellixia Publishing Limited in 2022 (www.ellixia.com)

978-1-716-03766-5

Acknowledgements: Most images and diagrams have been created by the writers and those from other sources are acknowledged with thanks for the permission to use them.

Written for

and the Welsh Government's educational Hwb.

Dr. Nick Winstone-Cooper studied chemistry and physics at Cardiff University before completing postgraduate research in nuclear chemistry, focusing on the creation of radioactive complexes of macrocyclic phosphines for application as heart and bone imaging agents in cancer diagnosis.

He worked extensively in the United Kingdom, France, North America, Italy and the Republic of Korea before moving into education and education consultancy.

Jasmine Tripconey studied Chemistry and Drug Discovery at the University of Bath. Her interests include the biochemistry of selective serotonin reuptake inhibitors and bioinorganic chemistry.

Dedicated

to

Mike Turbervill

A brilliant teacher, mentor, colleague and friend

Table of Contents

Introduction

General Introduction

This is the seventh in a series of chemistry handbooks which focus on the instrumental analysis of important chemical substances as an interesting way to develop an ability to interpret the infrared, mass and ^{1}H and ^{13}C NMR spectra of organic compounds. The handbooks are intended to bridge A-level and first year undergraduate chemistry courses and the molecules have been selected carefully to be appropriate to the nature of what students are expected to understand. Too often, students are introduced to chemical compounds which can seem to have been selected at random and so, in these handbooks, the application of the molecules is explained. This handbook concentrates on the analysis of fifteen selected amino acids out of the twenty natural compounds.

Each of the chapters follows the same style with an introductory page discussing the application and history of the development of the medicine with a summary of the elemental data and the formula mass, leading to the determination of the empirical and molecular formulas of the compound.

The following sections commence with examination and interpretation of the infrared spectrum. Until the introduction of routine multinuclear NMR spectroscopy, infrared (IR) spectroscopy, together with mass spectrometry was one of the most important analytical techniques available. IR spectroscopy is still of fundamental importance as it provides evidence for the presence or absence of particular functional groups but it is important to remember that no IR spectrum can quantify the number of any type of functional group. The spectrum does, however, give clues to the possible structures of the molecule. When there are a number of feasible isomers, however, the IR spectrum rarely allows us to differentiate between them.

Having interpreted the IR spectrum which can give important clues to the structure of the molecule we move on to examining the mass spectrum of the compound. All of the spectra are electron ionisation mass spectra. The absence of fragments predicted from the possible structures means nothing since, even though the travel of the ionised fragments is very brief, they can fragment on their journey. The most important and well known example of this is the McLafferty rearrangement of carbonyl, $C = O$, containing molecules.

The presence of certain fragments is, however, of profound importance and the most notable example is the peak at m/z = 77 which is only ever due to a benzene ring with one substituent, C_6H_5-. Similarly, peaks at m/z = 76 which is usually assignable to a benzene ring with two substituents. There are three isomers of disubstituted benzene rings, denoted as ortho $-$, meta $-$ and para $-$, and it is rare that the mass spectrum will determine which isomer is present. That is where multi-nuclear nuclear magnetic resonance (NMR) spectroscopy demonstrates its overwhelming significance.

NMR spectra contain three important measurements: the chemical shift, the multiplicity of the peaks and the integrals of the peaks. The ^{13}C NMR spectra only ever contain singlets due to the fact it is only ^{13}C isotopes which resonate and since they are much rarer than ^{12}C isotopes (comprising only about 1.1% of all stable carbon isotopes) there is an extremely low likelihood of two being adjacent in a molecule. If they did occur then there would be coupling and hence the signals would be multiplets.

Nevertheless the ^{13}C NMR spectrum is extremely important since the range of chemical shifts is large (δ 0 – 220 ppm) and the regions are clearly differentiated. This permits conclusions to be drawn and, as we shall see, they are often a better starting point than the ^1H NMR spectra and provide evidence for the presence of particular functional groups.

The ^1H NMR spectra contain three important different facets: the chemical shift identifies the type of environment occupied by the hydrogen atoms, the integral informs us of the proportion of all the hydrogen atoms that are chemically and magnetically equivalent whilst the multiplicity indicates the number of hydrogen atoms bonded to adjacent carbon atoms.

Together, the ^1H and ^{13}C NMR spectra provide exceptional evidence for the structure of the molecule but this also depends on knowledge of the molecular formula and evidence from the infrared and the mass spectra. If any of the spectra contradicts a proposed structure then the structure must be wrong and it is then time to start again.

Each chapter concludes with display of the concluded structure and the molecule's systematic name.

We will now discuss the general properties of amino acids before examining the elemental data and spectra of fifteen of them to determine their structures.

The final part of this volume comprises pen portraits of a number of the chemists and biochemists who either isolated the amino acids considered in this volume or deduced the structures of the compounds or achieved both.

Introduction to Amino Acids

Amino acids are important because they are the monomers used by nature to produce proteins, now properly known as polypeptides. Biochemists and physiologists regard them as one of three classes of compounds which, along with monosaccharides and triglycerides, as *the building blocks of life*.

There are approximately five hundred known amino acids but only twenty naturally occurring amino acids are used to manufacture protein. Nine of these are known as essential amino acids as we have to acquire them from foods, whether cheese, plants or meats. The rest are produced by our bodies.

The first few amino acids were discovered in the early 1800s whilst the last one to be isolated and characterised was in 1935. The history of the discoveries is briefly described in the introduction to each amino acid's chapter.

All acyclic amino acids i.e. those without a ring have the same basic structure:

where the side chain R, can be an alkyl group, a hydrogen atom or other groups containing sulfur or oxygen.

The general structure is important for many reasons but there are three important matters to note:

- The carboxylic acid ($-CO_2H$) and the amino (usually $-NH_2$) groups are terminal i.e. attached to the same carbon atom. Since the $-CO_2H$ group is acidic and the amino group is basic this has significant consequences which can be used to explain the fact that most amino acids are colourless crystalline solids which are soluble in water but usually insoluble in organic solvents.

- With the exception of glycine, which contains two hydrogen atoms, all amino acids are chiral. Chiral molecules exist in at least two forms which rotate the plane of polarised light in different directions but by the same angle. Chirality and optical activity is explained in full in Volume 1 of this series.
  When prepared under laboratory conditions all but glycine (which is achiral) are produced as a racemic mixture of the two optical isomers i.e. 50% of the molecules will rotate plane polarised light to the left and 50% will rotate plane polarised light to the right. The number of possible isomers equates to 2^n where n is the number of chiral carbon atoms in the molecule.

  It is fascinating, however, that all chiral amino acids produced in nature are so called left handed molecules meaning that they rotate plane polarised light to the left. This is believed to be due to the enzymes which produce the amino acids also being chiral. There are exceptions to this including L – alanine. L – alanine is a non-essential amino acid but its mirror image, D-alanine, is crucial in the construction of cell walls and is also found in some molluscs and mammalian muscle.

- The central carbon atom is known as the alpha (α) carbon and in a physiological, aqueous solution both the carboxylic acid and amino groups become ionised producing $-NH_3^+$ and $-COO^-$ groups. This is important because it is this property that enables the amino acids to bond to produce proteins, which are essentially polymers of amino acids and which may comprise many thousands of amino acid monomers.

Physical and Chemical Characteristics of Amino Acids

The number of amino acids in a protein and their sequence determines the shape, size and function of a protein. This is largely due to the side chains and the amino acids are bonded through an elimination reaction where the carboxylic acid and the amino acid groups produce a peptide bond by *elimination* of a water molecule as shown.

Such reactions are also known as dehydration reactions but this is an older name which is rarely used now. The equation above shows the overall reaction but the mechanism involves $-CO_2^-$ and NH_3^+ groups which are discussed below.

Polypeptides

Polypeptides (proteins) contain many thousands of amino acids bonded through this type of peptide chain. Each polypeptide has a free amino acid at one end, known as the terminal N group, whilst the other end possesses a carboxylic acid group and is known as both the *C-terminus* and the *C-terminal tail*.

Melting Points

For very small molecules, amino acids have surprisingly high melting points which are usually between 200 – 300°C. The boiling points are hard to determine because they tend to decompose.

The reason for the high melting points is easy to understand when we know their molecular structures and their stereochemistry. In the crystalline solid there is a transfer of a hydrogen ion from the carboxylic group to the amino groups. This creates a substance with both a *basic, amino*, group $-NH_3^+$ and an acidic, carboxylic acid group, $-CO_2^-$ as shown below:

This type of substance, which is overall neutral, is known as a *zwitterion* and leads the molecules to be bonded tightly, ionically, together. To separate the molecules requires overcoming these strong intermolecular forces which requires considerable energy and this explains why they usually have surprisingly high melting points.

Solubility

It is generally found that amino acids are soluble in water and polar solvents but insoluble in non-polar organic solvents. This, again, is due to the presence of zwitterions where the ionic attractions between the ions are replaced by polar interactions with polar water molecules. The absence of polarity in non-polar organic solvents explains the lack of solubility of amino acids in non-polar solvents. As with any other organic compound the solubility in water decreases with increasing length of the side, 'R' chain.

Part I

Elemental data and correlation charts

The following pages present:-

- A truncated Periodic Table showing the first four periods of the elements; which include all those elements involved in the compounds discussed in this series;
- A correlation chart for the analysis of infra red spectra;
- A concise summary of assignable fragments in the electron ionisation (EI) mass spectrometry analysis. Although there are a variety of ionisation techniques we only consider the simplest and, most commonly used, technique.
- Correlation charts and coupling constants for ^{1}H and ^{13}C nmr spectroscopy;

The Periodic Table of the Elements

Key
atomic number
Symbol
name
relative atomic mass

(1)	(2)		(3)	(4)	(5)	(6)	(7)	(0)
1 **H** hydrogen 1.0								**2** **He** helium 4.0
3 **Li** lithium 6.9	4 **Be** beryllium 9.0		5 **B** boron 10.8	6 **C** carbon 12.0	7 **N** nitrogen 14.0	8 **O** oxygen 16.0	9 **F** fluorine 19.0	10 **Ne** neon 20.2
11 **Na** sodium 23.0	12 **Mg** magnesium 24.3		13 **Al** aluminium 27.0	14 **Si** silicon 28.1	15 **P** phosphorus 31.0	16 **S** sulfur 32.1	17 **Cl** chlorine 35.5	18 **Ar** argon 39.9
19 **K** potassium 39.1	20 **Ca** calcium 40.1		31 **Ga** gallium 69.7	32 **Ge** germanium 72.6	33 **As** arsenic 74.9	34 **Se** selenium 79.0	35 **Br** bromine 79.9	36 **Kr** krypton 83.8

3	4	5	6	7	8	9	10	11	12
21 **Sc** scandium 45.0	22 **Ti** titanium 47.9	23 **V** vanadium 50.9	24 **Cr** chromium 52.0	25 **Mn** manganese 54.9	26 **Fe** iron 55.8	27 **Co** cobalt 58.9	28 **Ni** nickel 58.7	29 **Cu** copper 63.5	30 **Zn** zinc 65.4

Infrared Correlation Chart

Bond	Functional group	Wavenumber (cm⁻¹)
C – C	Alkanes and alkyl groups	750 – 1100
C – X	Haloalkanes (X = Cl, Br or I)	500 – 800
C – F	Fluoroalkanes	1000 – 1350
C – O	Alcohols, carboxylic acids and esters	1000 – 1300
Aliphatic C = C	Alkenes	~ 1650
C = O	Aldehydes, ketones, carboxylic acids, esters and acid chlorides	~ 1750

Bond	Functional group	Wavenumber (cm⁻¹)
Aromatic C = C	Aromatic compounds	1450 – 1650 (Multiple peaks)
C ≡ N	Nitriles	~ 2250
C – H	Alkyl groups, alkenes and aromatic compounds	2850 – 3000 (alkanes) 3000 – 3200 (alkenes and aromatics)
O – H	Carboxylic acids	2500 – 3500
N – H	Amines and amides	3300 – 3500
O – H	Alcholos and phenols	3200 – 3600

Notes: In the table above, Ar refers to an aromatic ring such as benzene whilst X refers to any of F, Cl, Br or I.

Some peaks are of little use for identification purposes since, for example, most organic compounds contain C – H bonds and so the presence of peaks just below 3000 cm⁻¹ is of little use for identification purposes. There is, however, a distinction between the C – H peaks above and below this wavenumber: aromatic compound C – H bonds appear above i.e. to the left of 3000 cm⁻¹ whilst aliphatic C – H bonds appear below, to the right, of 3000 cm⁻¹.

The combination of peaks is also important. For example, a carboxylic acid contains both a C =O and O – H bond so the presence of both is necessary for the confirmation of this class of compound.

Scale (cm⁻¹): 3600 3200 3000 2800 2600 2400 2200 2000 1800 1600 1400 1200 1000 800 600 400

Diagram bands: RO – H; N – H; =C – H; ArC – H; –C – H; RCOO – H (carboxylic acids); C ≡ N; C = N; C = C; C = O; Ar – H; C – O; C – C; C – X

Common Mass Fragmentation Ions

The following table shows a large number of fragmentation assignments which are relevant to the molecules in this volume. There are a number of important points to note:-

- If a molecule contains a *chlorine* atom then that fragment will exhibit two peaks, two units apart, due to the existence of the ^{35}Cl and ^{37}Cl isotopes. The heights of these peaks will be in the proportion 3:1 and there will always be two other peaks at m/z = 35 and 37, also in the ratio 3:1 due to the natural occurrence of the isotopes. If any of these four peaks are absent then chlorine is not present. The presence of chlorine will already, however, be recorded in the elemental composition of the compound.

- Similarly to chlorine, if a molecule contains a *bromine* atom then there will be two peaks for the molecular fragment due to the existence of ^{79}Br and ^{81}Br. Since these two isotopes exist in nearly equal proportions then the peak heights of these fragments will be of approximately equal height and there will, of course, also be peaks of equal height at m/z 79 and 81.

- *Aromatic* compounds such as those containing a benzene ring will usually exhibit a peak at m/z = 77 due to the existence of the C_6H_5- functional group. If there is a peak at this m/z ratio then it is almost always due to this. There will then also be peaks of lower m/z value due to fragmentation of the ring but these can also be due to the fragmentation of aliphatic chains and so the clue is in the m/z = 77 peak. More highly substituted benzene rings will exhibit peaks at m/z = 76, 75 etc; The existence of the aromatic portion of the molecule is, however, also and *always* conclusively demonstrated by the ^1H and ^{13}C nmr spectra. In the following table, the aromatic fragments are indicated by [a].

- Many mass spectrometry measurements may be completed in little more than one second. Whilst in everyday life one second is very brief it is, in physical terms, quite long. This means that the unstable ionised molecule may fragment or rearrange itself on its journey along the apparatus resulting in peaks that would not be predicted simply by considering the ripping apart of a molecule. peaks in the table below which result from rearrangement of an ionised molecule or fragments are indicated by [b] after the fragment's formula.

- Some molecules may produce different fragments of the same m/z and are listed as bullet points in the table below.

Table of mass fragments

m/z	Ion	m/z	Ion	
15	$[CH_3]^+$	65	$[C_5H_5]^{+a}$	
17	$[OH]^+$	67	$[C_5H_7]^+$	
18	$[H_2O]^+$	69	$[C_5H_9]^+$	
26	$[CN]^+$	70	$[C_5H_{10}]^+$ or $[C_4H_6O]^+$	
27	$[C_2H_3]^+$	71	• $[C_5H_{11}]^+$ • $[C_3H_7\text{-}C{=}O]^+$	
28	$[C_2H_4]^+$	72	$[C_2H_5\text{-}CO\text{-}CH_2{+}H]^b$	
29	$[C_2H_5]^+$ $[CHO]^+$	73	• $[C_3H_7OCH_2]^+$ • $[C_2H_5O\text{-}C{=}O]^+$	• $[C_3H_7CHOH]^+$ • $[C_2H_5OCHCH_3]^+$

m/z	Ion	m/z	Ion
30	$[CH_2NH_2]^+$	74	$[CH_2\text{-}COOCH_3+H]^{+b}$
31	$[CH_2OH]^+$ $[OCH_3]^+$	75	$[C_2H_5O\text{-}C{=}O+2H]^{+b}$ $[C_2H_5COO+2H]^{+b}$
35 & 37	$[^{35}Cl]^+$ & $[^{37}Cl]^+$	77	$[C_6H_5]^{+\,a}$
39	$[C_3H_3]^{+\,a}$	78	$[C_6H_5+H]^{+\,ab}$
40	$[CH_2CN]^+$	79	$[C_6H_5+2H]^{+\,ab}$ $[^{79}Br]^+$
41	$[C_3H_5]^+$ $[CH_2CN+H]^{+b}$	81	$[C_6H_9]^+$ $[^{81}Br]^+$
42	$[C_3H_6]^+$	82	$[C_6H_{10}]^+$ $[C^{35}Cl^{35}Cl]^+$
43	$[C_3H_7]^+$ $[CH_3C{=}O]^+$	83	$[C_6H_{11}]^+$ $[CHCl_2]^+$ (also 85&87)
44	$[CH_3CH\text{-}NH_2]^+$	84	$[C_6H_{12}]^+$ $[C^{35}Cl^{37}Cl]^+$
45	$[CH_3CHOH]^+$ $[CH_2OCH_3]^+$ $[CH_2CH_2OH]^+$ $[COOH]^+$	85	$[C_6H_{13}]^+$ $[C_4H_9\text{-}C{=}O]^+$
49	$[CH_2^{35}Cl]^+$	86	$[C_3H_7\text{-}CO\text{-}CH_2+H]^{+b}$ $[C^{37}Cl^{37}Cl]^+$
50	$[C_4H_2]^{+\,a}$	88	$[CH_2\text{-}COOC_2H_5+H]^{+b}$
51	$[CH_2^{37}Cl]^+$ $[C_4H_3]^{+\,a}$	89	$[C_3H_7\text{-}O\text{-}C{=}O+2H]^{+b}$ $[C_3H_7COO+2H]^{+b}$
52	$[C_4H_4]^{+\,a}$	90	$[C_6H_5\text{-}CH]^+$
53	$[C_4H_5]^+$	91	$[C_6H_5\text{-}CH_2]^+$ $[C_6H_5\text{-}CH+H]^{+b}$
54	$[H_2CH_2CN]^+$ $[CH_3CHCN]^+$	92	$[C_6H_5\text{-}CH_2+H]^{+b}$
55	$[C_4H_7]^+$	93	$[C_7H_9]^+$ $[CH_2^{79}Br]^+$
56	$[C_4H_8]^+$	94	$[C_6H_5O+H]^+$
57	$[C_4H_9]^+$ $[C_2H_5\text{-}C{=}O]^+$	95	$[CH_2^{81}Br]^+$
58	$[CH_3\text{-}CO\text{-}CH_2+H]^{b}$	97	$[C_7H_{13}]^+$
59	$[C_2H_5OCH_2]^+$ $[CH_3O\text{-}C{=}O]^+$ $[C_2H_5CHOH]^+$ $[CH_3O\text{-}CHCH_3]^+$	105	$[C_6H_5C{=}O]^+$ $[C_6H_5\text{-}CH_2CH_2]^+$
60	$[CH_2\text{-}COOH+H]^{+b}$	107	$[C_6H_5\text{-}CH_2O]^+$
61	$[CH_3COO+2H]^{+b}$ $[CH_3OCO+2H]^{+b}$	108	$[C_6H_5\text{-}CH_2O+H]^{+b}$
63	$[C_5H_3]^{+a}$	119	$[C_6H_5\text{-}C(CH_3)_2]^+$

[a] Good diagnostics for benzene ring compounds.

[b] Where a fragment results from a rearrangement with the movement of *one* hydrogen atom from one carbon to another this is indicated by the fragment followed by +H.

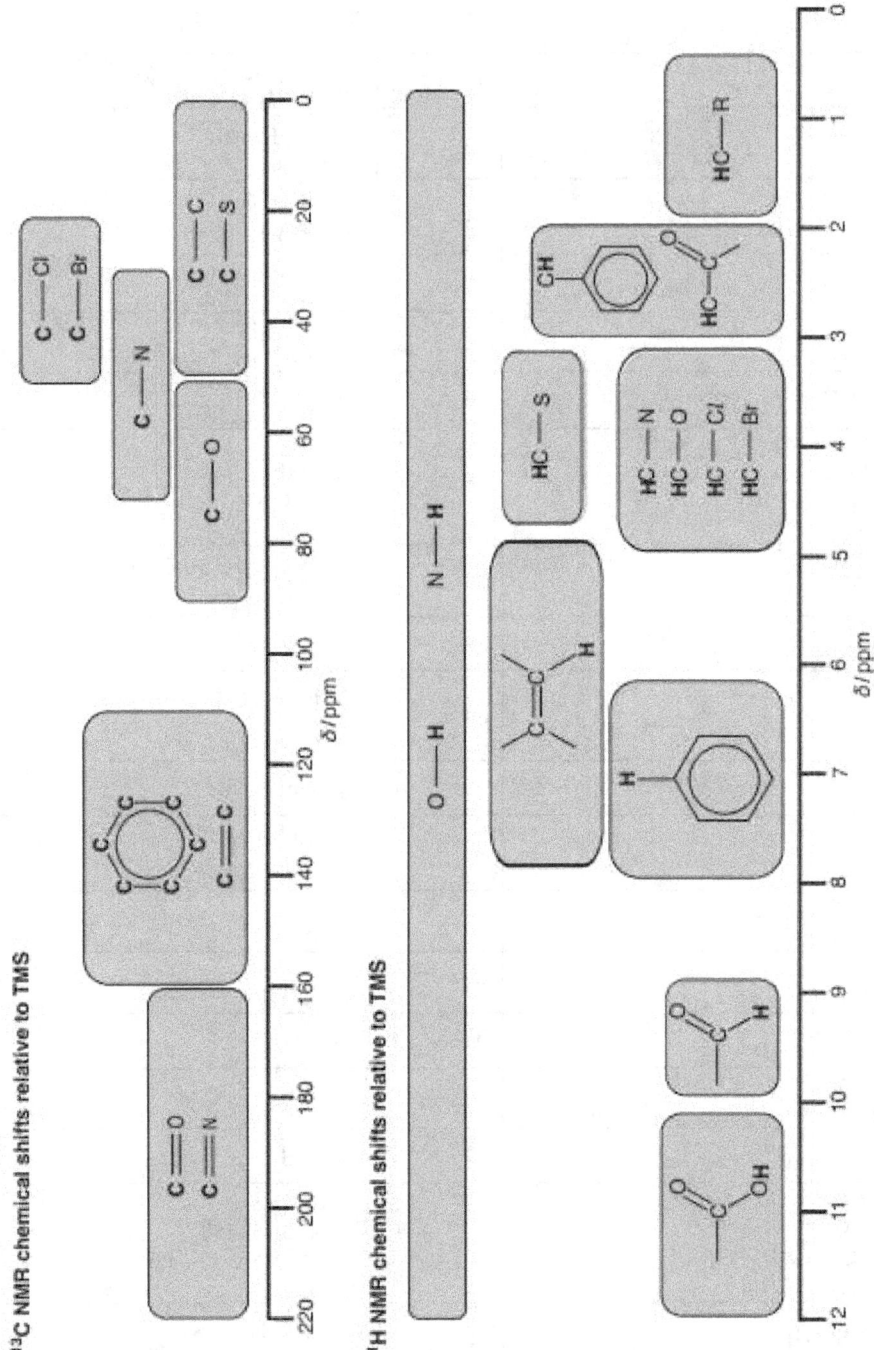

^{13}C NMR chemical shifts relative to TMS

^{1}H NMR chemical shifts relative to TMS

^{1}H NMR Coupling Constants

Aliphatic Alkenes

Isomerism	Coupling constant (J) range (Hz)
Geminal	0 – 5
Vicinal (cis) / Vicinal (Z –)	5 – 14
Vicinal (trans) / Vicinal (E –)	15 – 20

Substituted Aromatic Compounds

Designation	Ortho –	Meta –	Para –
Structural formula			
Coupling constant range (J):	7 – 10	2 – 3	0 – 2

Part II

Analysis of amino acids

Each of the following chapters is prefaced with an, unnumbered, introductory page which notes the history of the amino acid and its discover(s). The plant or animal sources of the amino acid are recorded together with its melting point and boiling point, both in ᵒC, and the angle it rotates plane polarised light.

Each introduction ends with the elemental composition, the formula mass and states the empirical and molecular formulas.

The volume is not organised alphabetically rather the amino acids are studied in the order of their simplicity. It is also important to remember that the amino acids are all the *laevorotary* (L −) isomer with the exception of glycine which is not optically active. The term *laevorotary* is usually abbreviated to *laevo*. The chiral carbon atom, the one which confers optical activity is identified with an asterisk *.

At the conclusion of each chapter the molecular structure is recorded together with the systematic name(s).

Chapter I

L - Alanine

First isolated from silk fibroin in 1879, L – alanine is unusual in that it had already been synthesised in 1850 by Adolph Strecker who reacted ethanal with ammonia in a reaction catalysed by hydrogen cyanide. Its name comes from the German for ethanal (acetaldehyde).

Due to its simple structure and ease of metabolic synthesis, L – alanine is considered to be one of the earliest amino acids produced in the genetic code. Biologically, it is synthesised from pyruvic acid (pictured below) which has in turn been produced by the breakdown of carbohydrates.

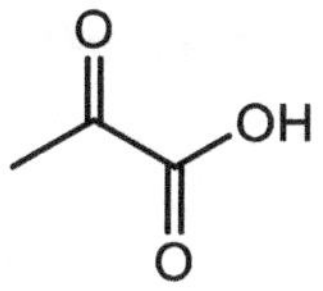

Pyruvic acid

L – alanine is also notable and unusual because its mirror image, D – alanine, is also found in small proportions in nature as it used to construct some cell walls and is also found in the tissues of molluscs and crustaceans as well as, in small amounts, mammalian muscle. D – alanine is also a component of the vitamin, pantothenic acid (below).

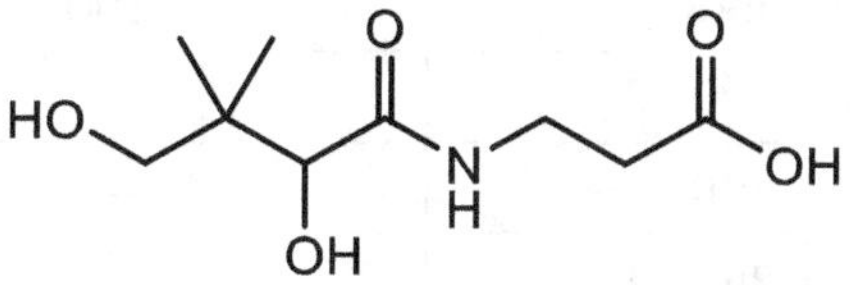

L - alanine has a melting point of about 300°C and decomposes at that temperature.

With a **formula mass** (M_r) of 89.09 gmol^{-1} , this compound has the **elemental composition**: C: 40.41%, H: 7.94%, N: 15.72%, O: 35.92%

This means that the empirical and molecular formulas are both $C_3H_7NO_2$.
Its angle of rotation is – 14.5° and has an aqueous solubility of 167.2 g dm^{-3} at 25°C.

L – Alanine

Infrared Spectrum

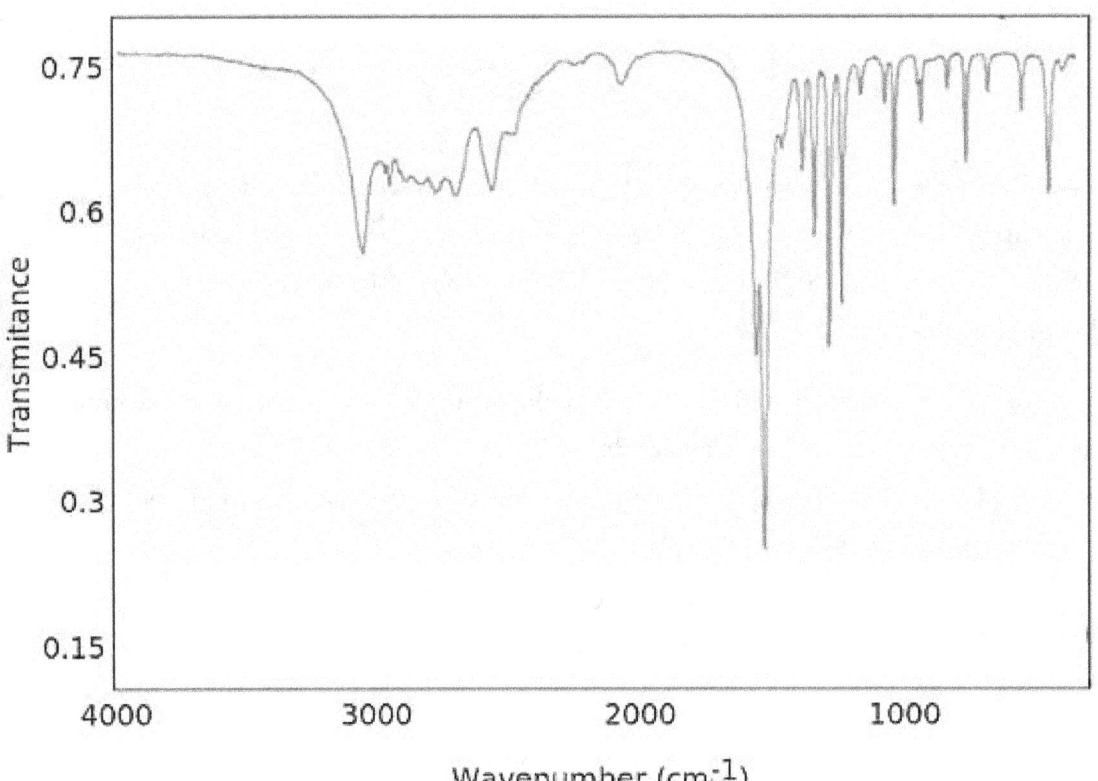

Wavenumber (cm⁻¹)

Observations

(√ / X)	Wavenumber range (cm⁻¹)	Wavenumber (cm⁻¹)	Assignment
?	3200 - 3700	3100	**O – H**
?	3200 - 3600	3100	**N – H**
X	3000 – 3300		**C – H (aromatic)**
√	2500 – 3000	3000 – 2500	**C – H (aliphatic)**
X	2200 – 2500		**C ≡ N**
√	1700 – 1800	1705	**C = O**
X	1700 – 1800		**C = N**
X	1600 – 1700		**C = C (aliphatic)**
X	1585 – 1600		**C – C (aromatic)**
X	1450 – 1600		**C – C (aromatic)**
X	1000 – 1300		**C – O**
X	700 – 1000		**C – X** (X = Cl, Br or I)

Conclusions

- The infrared spectrum demonstrates the presence of an N – H or O – H bond or both.

- There is evidence for a C = O functional group.

- There is no evidence for any aromaticity which, given the small size of the molecule, would be impossible.

L - Alanine

Mass Spectrum

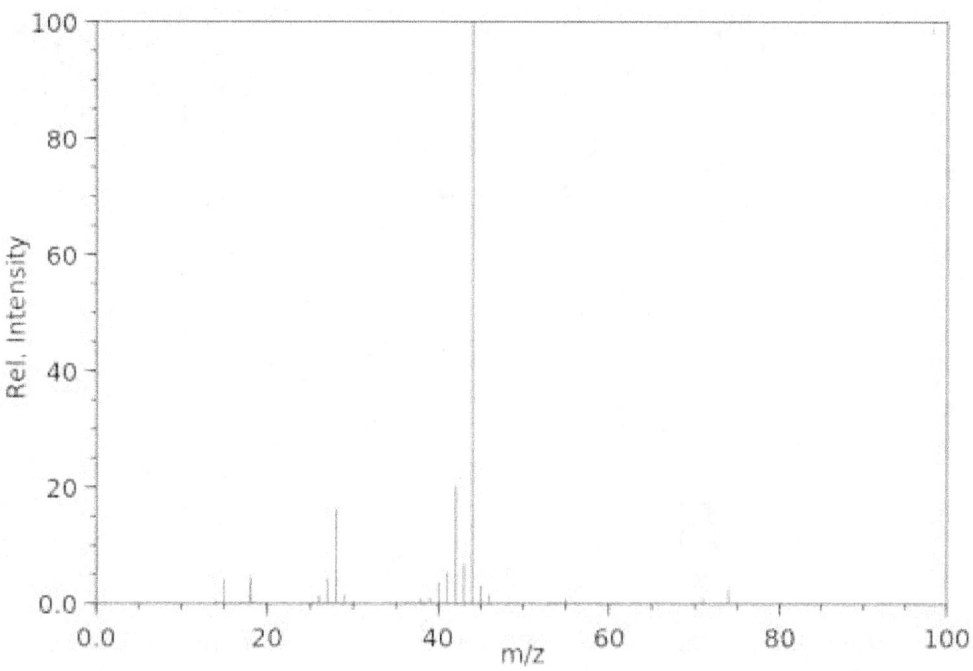

Observations

Charged fragments (m/z)	Assignment	Charged fragments (m/z)	Assignment
Molecular ion: 73	$[C_3H_7CHOH]^+$	Base peak: 44	$[CH_3CH-NH_2]^+$
45	$[CO_2H]_+$	27	$[C_2H_3]^+$ or $[CNH]^+$
42	$[C_3H_6]^+$	15	$[CH_3]^+$
28	$[C_2H_4]^+$ or $[CNH_2]^+$		

Conclusions

The mass spectrum confirms that there is a chain of carbon atoms and that there are both $-NH_2$ and $-CO_2H$ functional groups. This molecule has the molecular formula $C_3H_7NO_2$ and if one of the carbon and both oxygen atoms are used in the $-CO_2H$ and the nitrogen atom and another two hydrogen atoms are used in the amino group, totalling CH_3O_2N, this leaves us with only C_2H_4 to be assigned. Common sense tells us that this must be $-CH_2-CH_2-$ or $CH3-CH-$ and so there will be a terminal carboxylic acid, $-CO_2H$, group and a terminal amino, $-NH_2$ group somewhere on the C_2H_4 chain. No other structure is possible and so the structural formula must be:

It would be possible to put the amino group as a terminal group but the number of hydrogen atoms does not fit that possibility Even though we now have a structure we must still confirm that with the NMR spectra which is our next task.

NMR Spectra

The ¹H and ¹³C NMR spectra are displayed below:

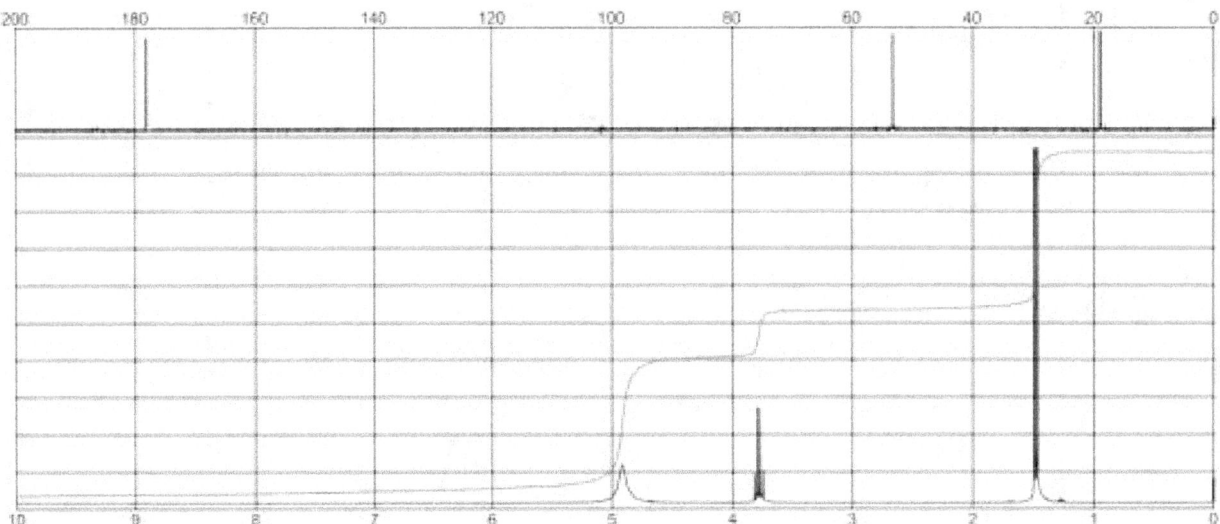

It is, however, more useful to predict the spectrum when there are one or more candidate molecules. which we will do first.

We only have one candidate molecule whose structure is repeated, for convenience, again below:

¹³C NMR Spectrum

If we examine the structure then we note that there should be three ¹³C peaks, all of equal integral.

From the data sheet we can see that they will occur in the following regions:

- δ 160 – 220 ppm due to the $- CO_2H$ carbon atom;
- δ 50 – 90 ppm due to the $- C - NH_2$ carbon atom;
- δ 0 – 50 ppm due to the $- CH_3$ carbon atom.

This is exactly what we observe but we must also confirm the structure with the ¹H NMR spectrum.

¹H NMR Spectrum

Predictions

If we examine the structure of the molecule again, we can predict the following

- There may or may not be a peak to the $- OH$ hydrogen atom. As explained in Volume 1 of this series, hydroxyl hydrogen atoms can be labile and, if they appear, then that can be anywhere but will be of integral one.

- There will be a peak of integral three due to the $- CH_3$ hydrogen atoms and this will appear somewhere between δ 0.5 and 2 ppm. This will be a doublet due to the presence of a single hydrogen atom on the adjacent carbon atom.

- There will be a quartet of integral one due to the hydrogen atom on the carbon attached to the $- CH_3$ and $- NH_2$ group. This will be a quartet since the $- NH_2$ group hydrogen atoms do not couple.

- Due to lability, there may or may not be a signal assignable to $- NH_2$ hydrogen atoms.

Observations

Upon examination of the actual spectrum we observe a peak at at δ 4.85ppm. This is due to water and can be ignored.

Other than the signal due to the hydroxyl hydrogen atom this is exactly what we observe as explained below.

The expanded ^1H nmr spectrum is shown below:

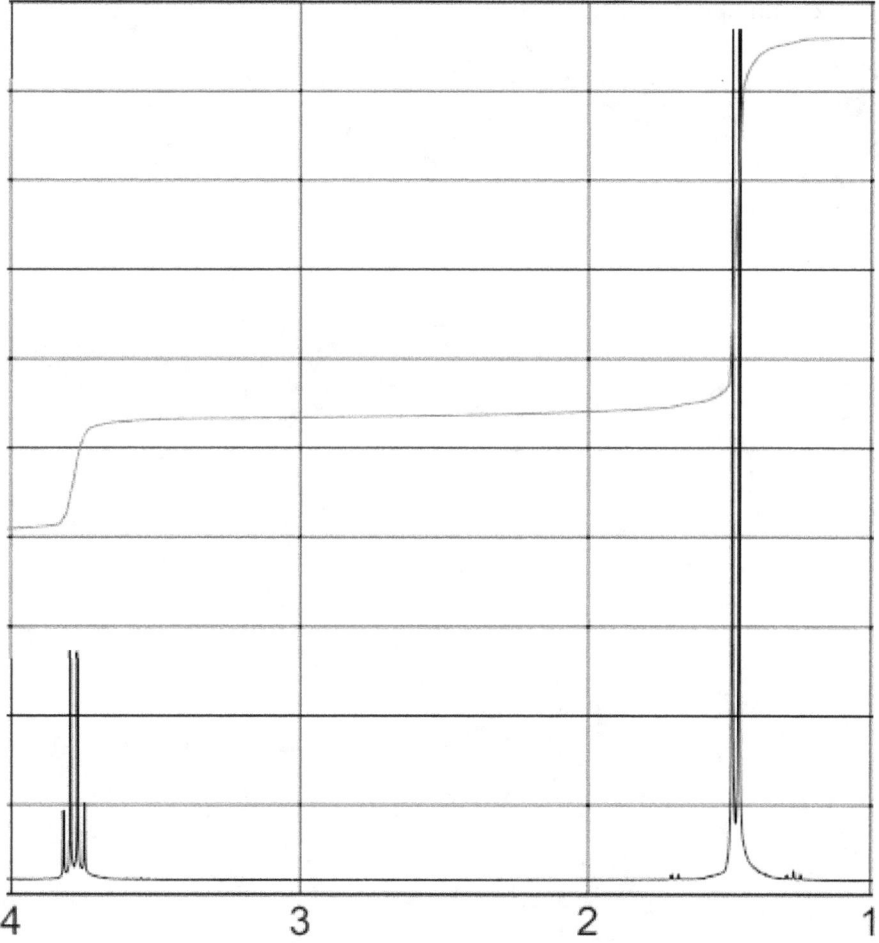

We observe:-

- A quartet of integral one at δ 3.75 ppm. This is a hydrogen atom bonded to a methyl, -CH_3, group;

- A doublet of integral three at δ 1.45 ppm which is assignable to that methyl group bonded to a carbon atom with one hydrogen atom attached to it.

Conclusions

Structure:

L – alanine has the following structure as we predicted and the chiral carbon atom is indicated, as is standard practice by an asterisk, *:

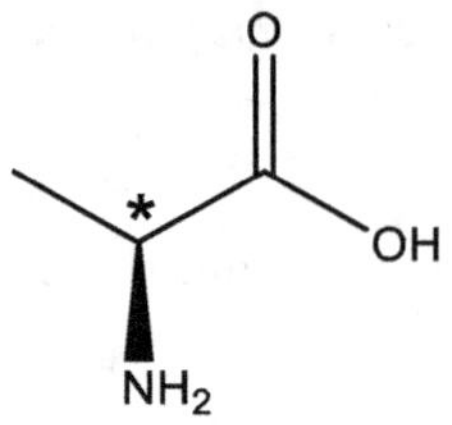

Systematic name: 2-aminopropanoic acid.

Chapter II

L – Cysteine

First isolated in 1810 by William Hyde Wollaston, L – cysteine is a non-essential amino acid and is one of the few sulfur-containing amino acids. It is essential for maintaining the structure of proteins and it is also a component of the antioxidant glutathione. The amino acid is used to produce another amino acid, L – taurine, a major constituent of bile and active in the large intestine. It is also important in collagen production and is found in beta-keratin which is the main protein in hair, skin and nails.

L – cysteine has been found to be essential for cardiovascular function as well as in the development of the central nervous system, skeletal muscle and the retina.

It is extremely unstable and, being nucleophilic, is readily oxidised. This is prevented by the formation of a dimer, cystine, where the molecules are connected by a disulfide bond.

L – cysteine has been found to have powerful antioxidant properties and it has also been marketed as an anti-ageing and a skin whitening substance.

Found in many high protein foods, cysteine is classified as a non-essential amino acid as it is produced by the metabolism of the amino acid, L – serine (Chapter IV). It is also, interestingly, essential for sheep to be able to produce wool and sheep must be given the amino acid in their feed.

L – cysteine has a melting point of 220°C, boiling point of ~300°C and, in a 5M HCl solution, rotates plane polarised light at an angle of +5°. Its solubility in water varies with pH and it is slightly soluble in ethanol. The variation of the aqueous solubility is highly significant and is addressed later in this chapter.

With a **formula mass** (M_r) of 121.16 g mol^{-1}, L – cysteine has the **elemental composition**: C: 29.72%, H: 5.84%, N: 11.56%, O: 26.41%, S: 26.41%

Empirical formula: $C_3H_7NO_2S$
Molecular formula: $C_3H_7NO_2S$

L – Cysteine

Infrared Spectrum

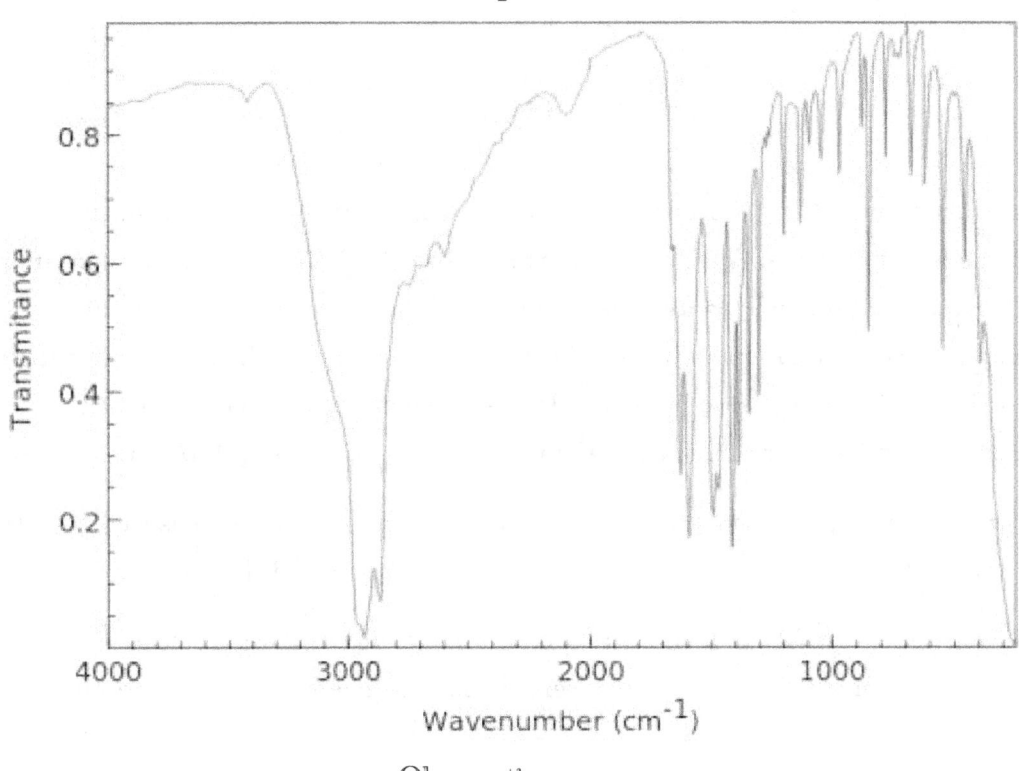

Observations

(√ /)	Wavenumber range (cm⁻¹)	Wavenumber (cm⁻¹)	Assignment
?	3200 - 3700	Broad peak merging with aliphatic C – H stretches	**O – H**
?	3200 - 3600	Broad peak merging with aliphatic C – H stretches	**N – H**
X	3000 – 3300		**C – H (aromatic)**
√	2500 – 3000	2920, 2860	**C – H (aliphatic)**
X	2200 – 2500		**C ≡ N**
?	1700 – 1800	Possible	**C = O**
?	1700 – 1800	Possible	**C = N**
?	1600 – 1700	Possible	**C = C (aliphatic)**
X	1585 – 1600		**C – C (aromatic)**
X	1450 – 1600		**C – C (aromatic)**
√	1000 – 1300		**C – O**
X	700 – 1000		**C – X (X = Cl, Br or I)**

Conclusions

This is a very complicated spectrum to interpret. The only thing we can say for certain is that the molecule is not aromatic. It is, in any event, too small for that but the broad peak which overlaps with the C – H stretch could be due to either or both of N – H and O – H as both nitrogen and hydrogen are present in the molecular formula.

Mass Spectrum

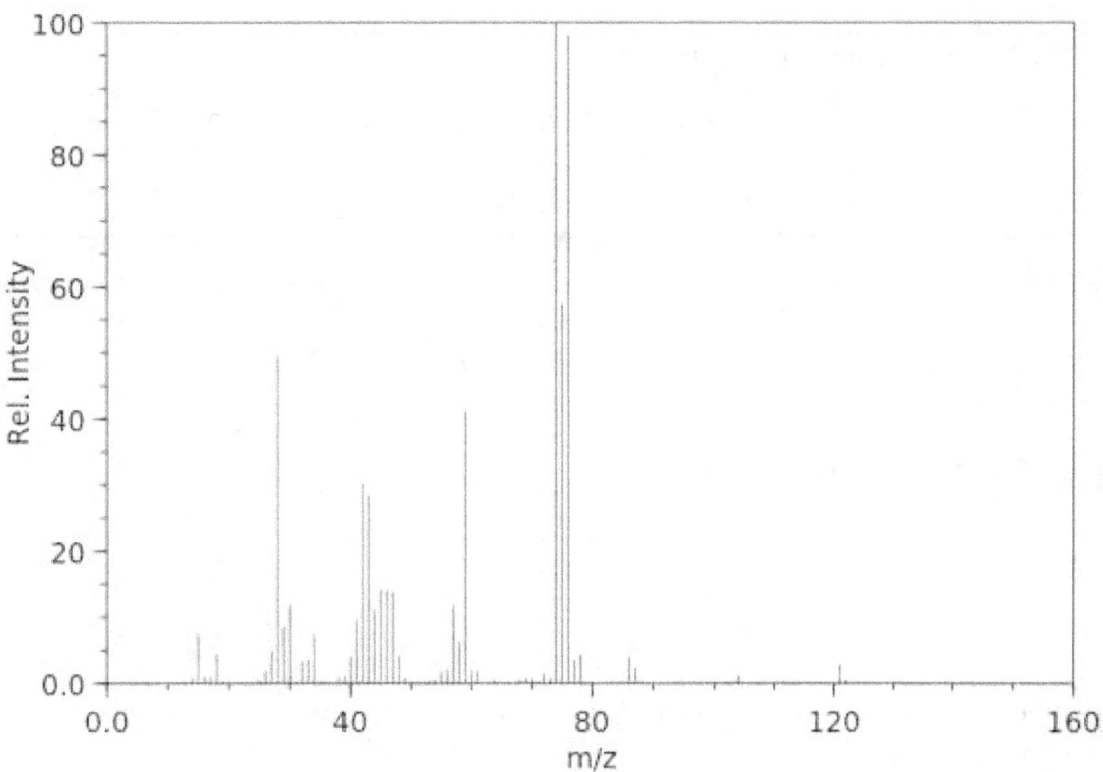

Observations

Charged fragments (m/z)	Assignment	Charged fragments (m/z)	Assignment
Molecular ion: 121	$[C_3H_7NO_2S]^+$	Base peak: 74	$[C_2H_4NO_2]^+$

88	$[C_3H_6NO_2]^+$	43	$[C_3H_7]^+$
76	$[C_2H_6NO_2]^+$	28	$[C_2H_4]^+$
59	$[C_3H_7O]^+$	15	$[CH_3]^+$

Conclusions

It is very hard to determine anything significant from this spectrum other than to propose that the molecule contains a carbon chain which, given the molecule's size is completely reasonable. If there is a chain then we have four possibilities:

I	II	III	IV

Our only hope is to examine the ^{1}H and ^{13}C NMR spectra.

NMR Spectra

The ¹H and ¹³C NMR spectra are displayed below:

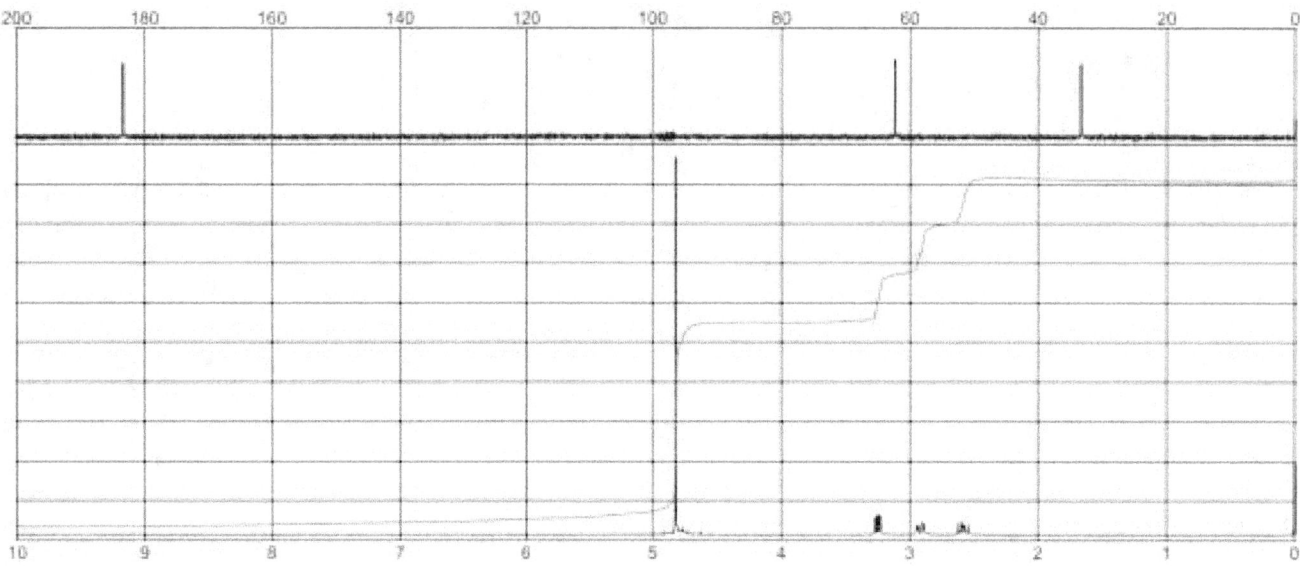

We will examine the ¹³C nmr spectrum first as that is very instructive and will then examine the expanded ¹H NMR spectrum.

¹³C NMR Spectrum

There are three peaks in the ¹³C NMR spectrum

- The peak at δ 183 ppm can only be due to a carboxylic acid carbon;
- That at δ 62 ppm is due to a C – N bonded carbon atom;
- The peak at δ 35 ppm is due to a C – S bonded carbon atom.

We can, therefore, conclude that the molecule does indeed contain a carboxylic acid (– CO_2H) and a carbon chain but candidates III and IV are not possible as there is no alkyl carbon signal.

Given the molecular formula, we can draw a partial structure as:

where the squiggles represent the remaining atoms amounting to NSH_3. in terms of numbers of the atoms of individual elements. These can only form – SH and – NH_2 groups but the question is which goes where? We have down to already seen that there are two possible isomers:

I   II

Fortunately we get a *Get Out Of Jail Free card* when determining which structure is correct.

- Amino acids always have a terminal amino group otherwise this molecule could not form an amide, formerly known as a peptide, bond;

- The amino nitrogen atom is tetrahedral due to its lone pair and so, due to stereochemistry, the – CH$_2$ – would be chemically and magnetically non – equivalent. In contrast, if the thio group was bonded to the terminal carbon atom then the – CH$_2$ – hydrogen atoms would experience free rotation and be chemically and magnetically equivalent.

- If the sulfur atom forms a bond with another molecule then its bonded hydrogen atom would have to go somewhere and there is nowhere it could go as the nitrogen atom could not accept a third hydrogen atom.

For complete confidence, however we must examine the ^{1}H nmr spectrum which is our final task.

<h2 style="text-align:center">^{1}H NMR Spectrum</h2>

If we examine and label the hydrogen atoms in the proposed structure we can make predictions about the multiplicity and integrals of the signal:

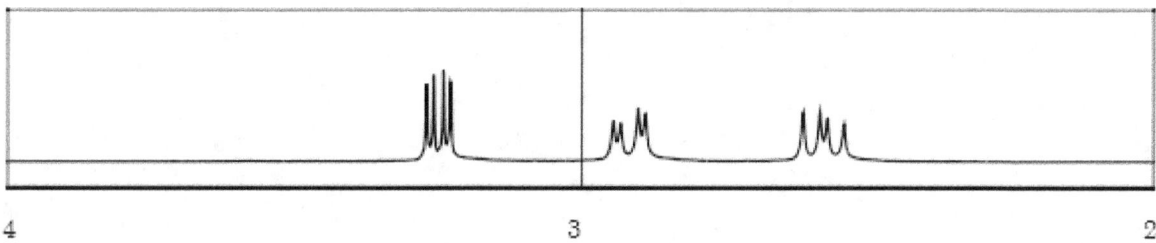

Due to the stereochemistry of the molecule the hydrogen atoms on the carbon atom adjacent to the thiol (–SH) group are not chemically and magnetically equivalent and so will produce separate signals but at similar chemical shift.

- H$_a$ will produce a singlet of integral one;
- The signal due to H$_b$ will be a doublet of doublets, of integral one, due to coupling with H$_{b'}$ and then with H$_c$;
- A similar signal, a doublet of doublets, of integral one, due to H$_{b'}$, will appear at a similar chemical shift to H$_b$ due to coupling with H$_b$ and with H$_c$;
- H$_c$ will produce a doublet of doublets of integral one due to coupling with H$_b$ and separately with H$_{b'}$.

This is exactly what we observe in the ^{1}H NMR spectrum as demonstrated by the expanded relevant portion of the spectrum which is displayed below (but without the integrals) :

The full spectrum (shown above) shows the integrals which are all one.

Conclusions

Structure:

Systematic name: L – cysteine and 2-amino-3-sulfhydrylpropanoic acid (IUPAC).

Chapter III

L – Glycine

Discovered in 1820 by the French chemist Henri Braconnot through the hydrolysis of gelatin, by boiling it with sulphuric acid, there were a number of other investigations and and discoveries but the most significant was by Jean-Baptiste Boussingault who showed that L – glycine also contains nitrogen.

L – glycine is used by the body, and is produced from L – serine (Chapter IV), to make proteins in the body but it is also a messenger for chemical signals in the brain. There have also been studies in its potential to treat schizophrenia and memory loss.

L – glycine has also been discovered through astronomical studies as present in a few meteorites but, most significantly, was detected in samples taken from the famous comet 67P/Churyumov–Gerasimenko.

A colourless and sweet tasting solid, L – glycine has a melting point of 233°C at which temperature it decomposes.

With a **formula mass** (M_r) of 75.07 gmol^{-1}, this compound has the **elemental composition**: C: 31.97%, H: 6.73%, N: 18.66%, O: 42.63% meaning that it has both empirical and molecular formulas of $C_2H_5NO_2$.

L – Glycine

Infrared Spectrum

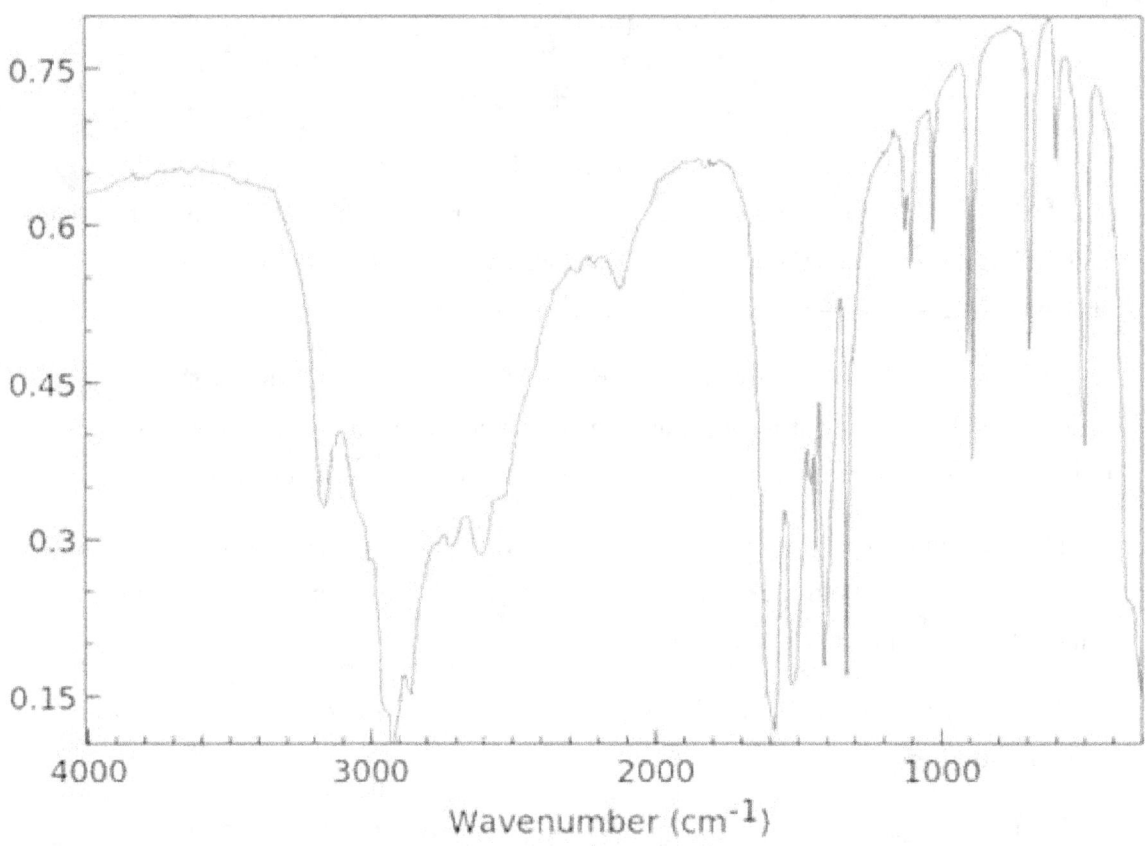

Wavenumber (cm^{-1})

(√ / X)	Wavenumber range (cm^{-1})	Wavenumber (cm^{-1})	Assignment
?	3200 - 3700	3180	O – H
?	3200 - 3600	3180	N – H
X	3000 – 3300		C – H (aromatic)
√	2500 – 3000	2920 – 2500	C – H (aliphatic)
X	2200 – 2500		C ≡ N
X	1700 – 1800		C = O
X	1700- 1800		C = N
?	1600 – 1700	1610	C = C (aliphatic)
X	1585 – 1600		C – C (aromatic)
X	1450 – 1600		C – C (aromatic)
√	1000 – 1300	1290	C – O
X	700 – 1000		C – X (X = Cl, Br or I)

Observations

This is a confusing spectrum since

* There might be – N–H and – O–H groups;
* The peak at 1610 cm^{-1} implies the presence of a C = C group but;
* The presence of a C – O at 1290 cm^{-1} suggests the presence of a carboxylic acid group.

In any event, the molecule is too small to have a C = C bond as there would be no room left for all the other atoms.

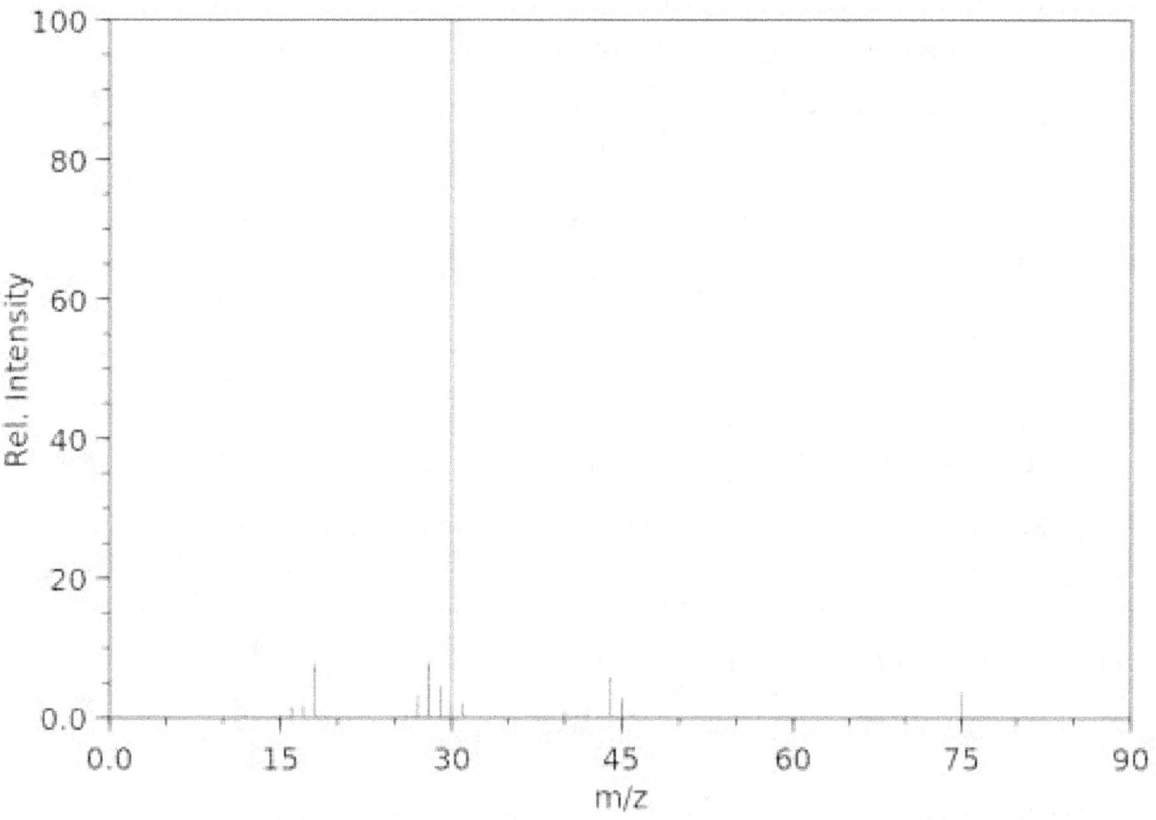

L – Glycine

Mass Spectrum

Observations

Charged fragments (m/z)	Assignment	Charged fragments (m/z)	Assignment
Molecular ion: 75	$[C_2H_5NO]^+$	Base peak: 30	$[CH_2NH_2]^+$

45	$[CO_2H]^+$	18	$[H_2O]^+$
29	$[C_2H_5]^+$	16	$[NH_2]^+$

Conclusions

There is no evidence for the presence of a C = C functional group / bond and the presence of both an amino group and a carboxylic acid group suggests that the only possible structure for such a small molecule is:

Due to insufficient numbers of hydrogen atoms it would not be possible for the molecule to possess the following structure:

In any event this molecule would be chiral and we know that glycine is not optically active.

We can learn much more from the ^1H and ^{13}C NMR spectra which is our last task

.NMR Spectra

The ¹H and ¹³C NMR spectra of the compound in ²H₂O (D₂O) are displayed below:

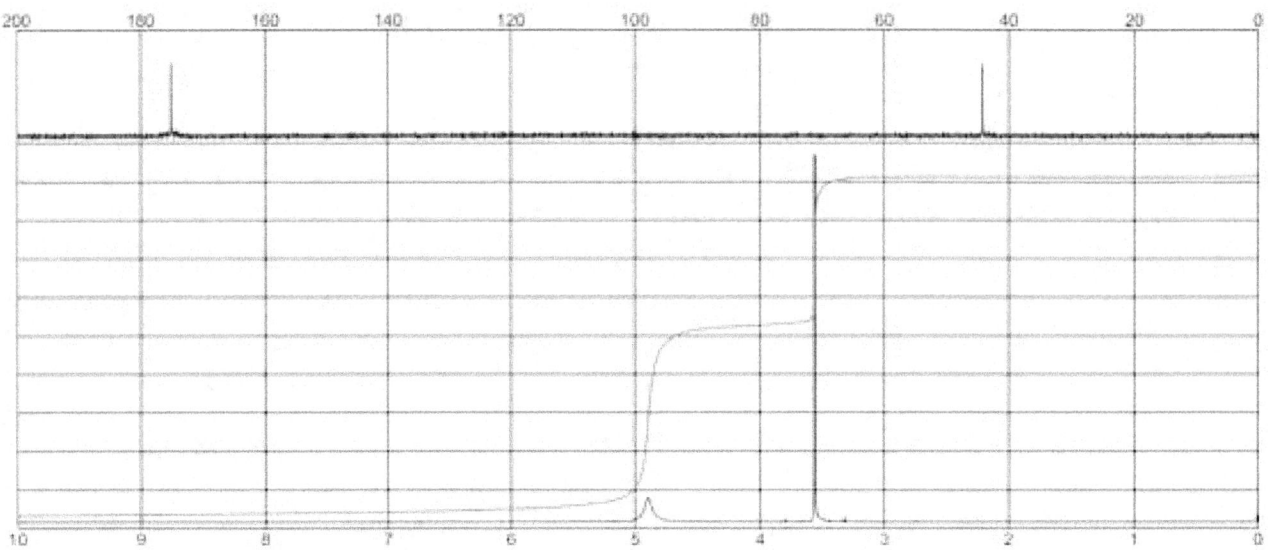

We will examine the ¹³C nmr spectrum first.

¹³C NMR Spectrum

There are two singlets each of integral one.

◼ The peak at δ 176 ppm is clearly due to a C = O carbon reinforcing the proposal that the molecule contains a carboxylic acid ($-$ CO₂H) group.

◼ The peak at δ 43 ppm can be assigned to a C $-$ N bond.

The only possible structure causing this spectrum is that predicted already:

and we can confirm or disprove this by examining the ¹H NMR spectrum.

¹H NMR Spectrum

It is wise to predict the ¹H NMR spectrum before examining the actual spectrum and we can predict the following :

◼ If it appears then H_a will produce a singlet of integral two anywhere in the entire region δ 0 $-$ 12 ppm. It will be a singlet because there is no coupling through the nitrogen atom;

◼ Again, if it appears, then H_c will produce a signal of integral one anywhere in the δ 0 $-$ 12 ppm region;

◼ H_b will produce a singlet in the region δ 3 $-$ 5 ppm. It will be a singlet because there are no hydrogen atoms on the carboxylic acid carbon atom and nitrogen in a C $-$ N bond doesn't couple.

Examining the actual ^{1}H NMR spectrum, as expected the hydroxyl hydrogen does not appear and there are two signals of relative integral 1:1 confirming the structure of the molecule is as predicted.

There are two signals but the peak at δ 4.85 ppm can be disregarded as that is well known to be due to ^{1}H impurities in solvents. This is now a good point to consider when signals do or do not appear.

Lability of amino and hydroxyl hydrogen atoms

One of the problems with analysing amino acids by ^{1}H and ^{13}C nmr spectroscopy is that most amino acids have very low, if any, solubility of organic solvents such as deuterated chloroform ($CDCl_3$ / C^1HCl_3). This restricts us to using heavy water (D_2O / ^1H_2O). Whilst demonstrating the importance of water in organic substances this then introduces the problem of the lability of amino and hydroxyl hydrogen atoms in aqueous solution.

In essence, labile hydrogen atoms will exchange rapidly with hydrogen atoms in water at such a speed that they may not be detectable in the one dimensional ^{1}H nmr spectroscopic analysis.

The reason for the lability of amino and hydroxyl hydrogen atoms is due to the fact that glycine (and many other amino acids) is zwitterionic in aqueous solutions which means that they can form cations or anions purely depending on the pH of the solution.

This is shown below, specifically for glycine, but the concept applies to many amino acids.

$$\overset{+}{H_3N}\diagup\!\!\diagdown CO_2H \underset{+H^+}{\overset{-H^+}{\rightleftharpoons}} \overset{+}{H_3N}\diagup\!\!\diagdown CO_2^- \underset{+H^+}{\overset{-H^+}{\rightleftharpoons}} H_2N\diagup\!\!\diagdown CO_2^-$$

In **acidic** solution   In **neutral** solution   In **alkaline** solution

There is rapid exchange with hydrogen atoms in the water molecules and this explains why many amino and hydroxyl hydrogen atoms are not detected.

It is well knows that ordinary water (^1H_2O) will produce a signal centred at δ 4.85 ppm. These signals are usually broad and ill – defined and can always be ignored.

In aqueous solution, glycine forms an acidic solution (pH ~2.3) and will largely exist as:

$$\overset{+}{H_3N}\diagup\!\!\diagdown CO_2H$$

The amino and hydroxyl hydrogen atoms will, though, exchange with the deuterated hydrogen atoms in the solvent at such a rapid rate that they appear invisible to the NMR spectrometer. This explains why there is only one signal in the ^{1}H nmr spectrum. The integral is irrelevant as they are relative to other signals which are, of course, absent in this spectrum

Conclusions

Structure:

NH₂ — CH₂ — COOH (aminoacetic acid structure)

Since there is no chiral carbon atom, i.e. one with four different substituents, this molecule is not optically active and is, in fact, the only achiral amino acid.

Systematic name: L – glycine and aminoacetic acid.

Chapter IV

L – Serine

L – serine appears naturally in proteins and is not essential to the human diet as it is produced by the biosynthesis of other amino acids such as glycine (Chapter III).

First extracted from silk, its name is derived from the Latin for silk (*sericum*). It is not essential to the human diet, since it is synthesised in the body from other metabolites although it is also found in eggs, meats, sardines and seaweed.

First isolated in 1865 by Emil Cramer, about whom relatively little is known, it has been found to be the precursor for other amino acids including not only L – glycine (Chapter II) but also L – cysteine (Chapter III).

Unusually for an amino acid, its mirror image, D – serine, is also important and it is found to be present in brain neurons and also acts as a signal in cartilage and the kidneys.

L – serine has a melting point of about 246°C when it decomposes and rotates plane polarised light, in aqueous solution, by -7°. It has a sweet flavouring is but also has minor umami and sour tastes at high concentration whilst D – serine has a musty but also sweet taste.

With a **formula mass** (M_r) of 105.09 g mol^{-1} , this compound has the **elemental composition**: C: 34.26 %, H: 6.73 %, N: 13.33 %, O: 45.68 % which means that it has identical empirical and molecular formulas of $C_3H_7NO_3$

L - Serine

Infrared Spectrum

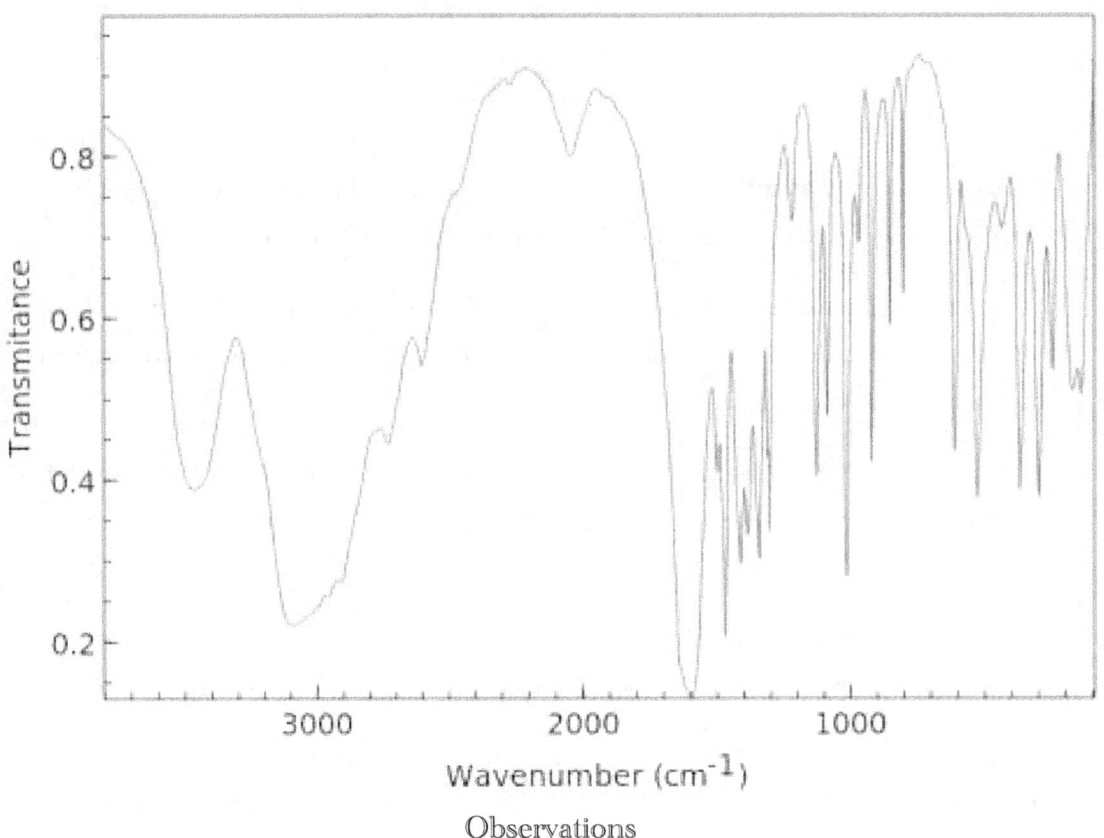

Observations

(√ / X)	Wavenumber range (cm⁻¹)	Wavenumber (cm⁻¹)	Assignment
?	3200 - 3700	3500 / 3100	O – H
?	3200 - 3600	3500 / 3100	N – H
X	3000 – 3300		C – H (aromatic)
√	2500 – 3000	3100 – 2900	C – H (aliphatic)
X	2200 – 2500		C ≡ N
X	1700 – 1800		C = O
X	1700 – 1800		C = N
√	1600 – 1700	1600	C = C (aliphatic)
X	1585 – 1600		C – C (aromatic)
X	1450 – 1600		C – C (aromatic)
√	1000 – 1300	Fingerprint region	C – O
X	700 – 1000		C – X (X = Cl, Br or I)

Conclusions

- This molecule has the molecular formula $C_3H_7NO_3$ and so can contain either or both N – H and O – H groups.

- Although there is a broad peak (3100 – 2900 cm⁻¹) the molecule is too small to be aromatic and the part of the stretch to the left might be due to an O – H or N – H bond.

- The peak at 1600 cm⁻¹ can be assigned to a C = C group.

Mass Spectrum

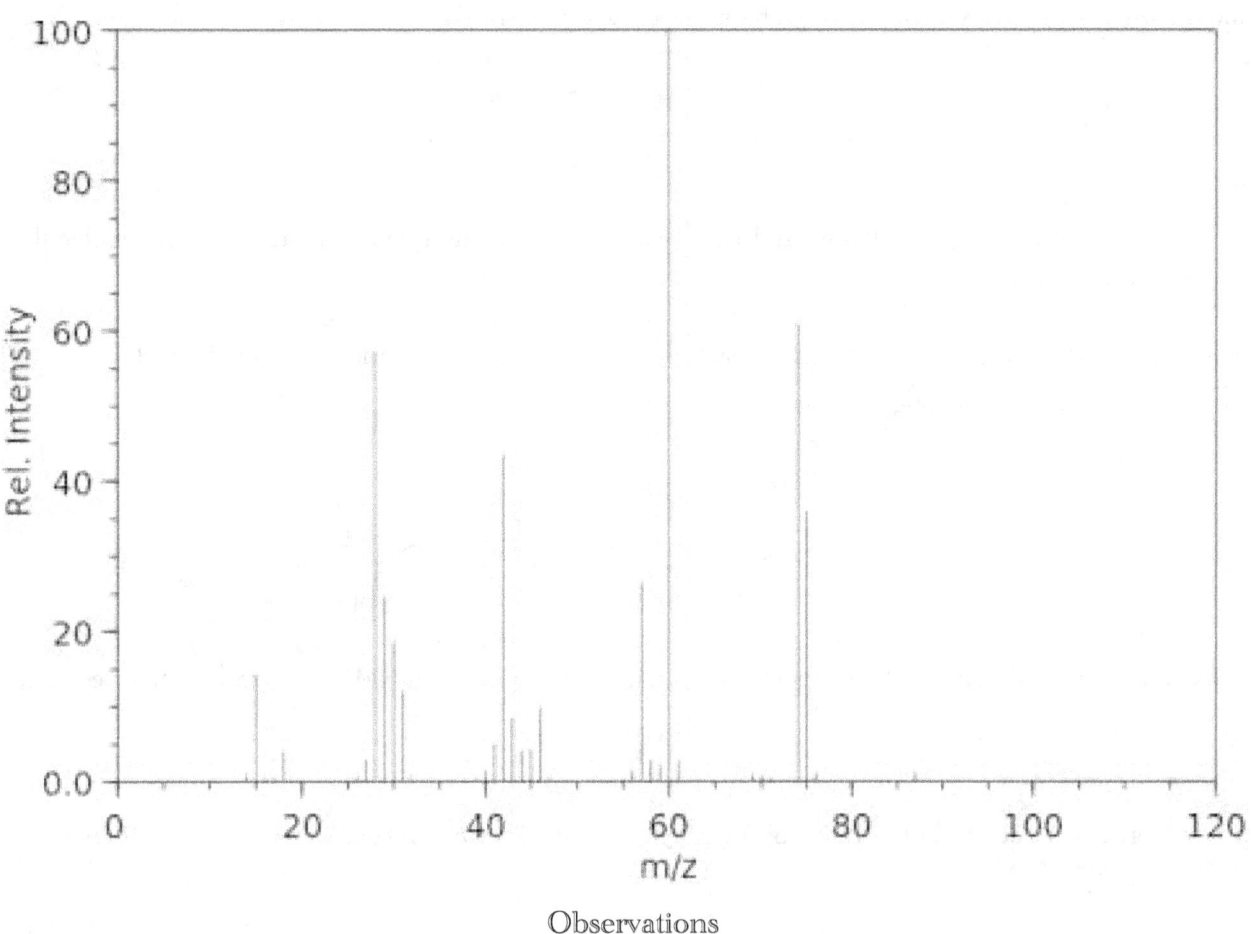

Observations

Charged fragments (m/z)	Assignment	Charged fragments (m/z)	Assignment
Molecular ion: 75	$[C_2H_5CO_2 + 2H]^+$	Base peak: 60	$[C_2H_6NO]^+$
57	$[C_2H_5C=O]^+$	42	$[C_3H_6]^+$
46	$[C_2H_6O]^+$	28	$[C_2H_4]^+$
45	$[CO_2H]^+$	15	$[CH_3]^+$

Conclusions

This is a fascinating spectrum as,

- It shows the presence of a carboxylic acid, $- CO_2H$, group which is not demonstrated immediately by the infrared spectrum until we accept that the correlation chart is a guide and not exact. This means that the peak we assigned to a C = C bond is actually due to a C = O functional group.

- It also suggests the presence of a three carbon chain.
  There are only three carbon atoms in the molecular formula. A carboxylic acid group must be at the end of a chain or substituted on to a chain. Since the molecule is small there is little likelihood of a substituent containing carbon as there are simply not enough carbon atoms for this structure.

As there are only three carbon atoms in the molecule and one of the carbon atoms is in the $-CO_2H$ group this means that, as a first approximation, the molecule's structure is:

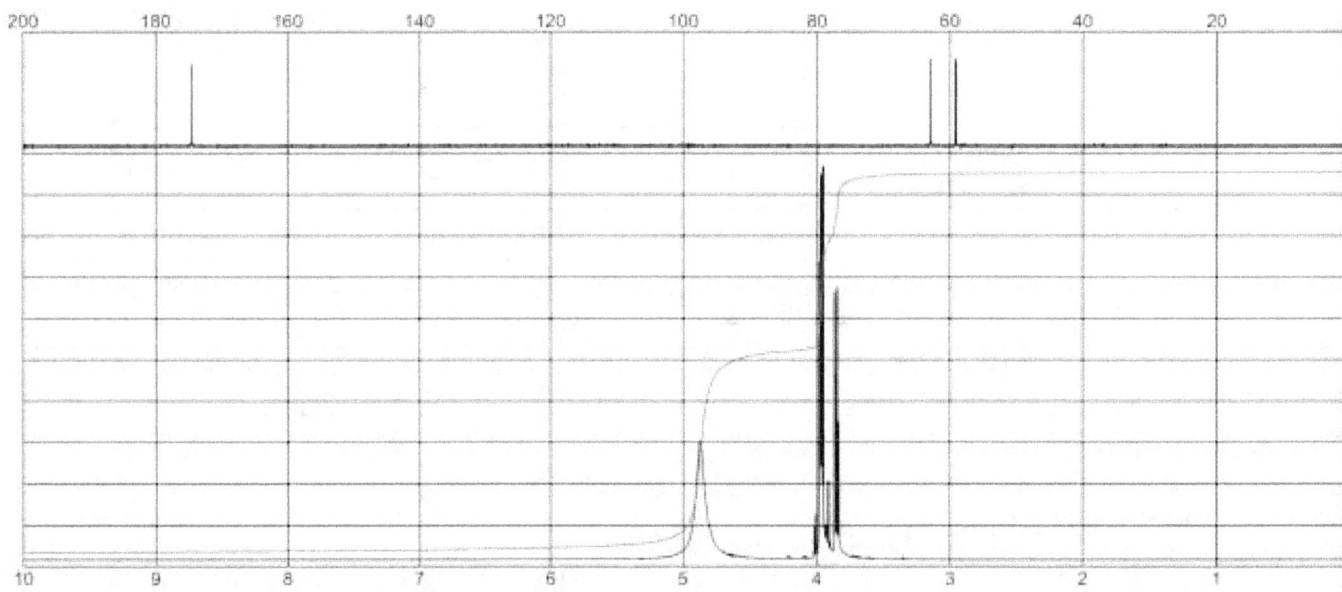

where the two functional groups R_1 and R_2 are comprised of the remaining atoms in the molecular formula which amount to H_3NO.

The obvious solution is that these must form an amino, $-NH_2$, group and a hydroxyl, $-OH$, group. This gives us two possible isomers for L - serine:

We can distinguish between these two isomers using the ^{13}C and ^1H NMR spectra which we consider next.

NMR Spectra

The spectra, recorded in D_2O (^2H_2O) are displayed below and we will consider the ^{13}C NMR spectrum first:

^{13}C NMR Spectrum

If either structure is correct then in both cases there will be a singlet, of integral one, in the region δ 160 – 220 ppm. There is a peak at δ 175 ppm which confirms the presence of a $-CO_2H$ carbon atom and further supports the proposed basic structure. We must, therefore, consider the consequences of the relative positions of the $- NH_2$ and $- OH$ groups.

We can distinguish between them by analysing the chemical shifts of the other two peaks in relation to the two possible candidates whilst continually bearing in mind that we might be completely wrong.

Candidate I

The substituted carbon atom attached to the – OH group will be deshielded by its proximity to so many oxygen atoms due to the greater electronegativity of oxygen whilst the carbon atom bonded to the amino, – NH_2, group will be not be really affected by the presence of the terminal carbon atoms at the other end of the molecule. This means that there will be significant separation between the two peaks:

The HO – **C** signal will appear at the higher end of the δ 50 – 90 ppm region whilst the H_2N – **C** will appear at the lower end of the δ 30 – 70 ppm and so there will be a significant separation between the two signals.

Candidate II

The terminal **C** – OH carbon atom will be unaffected by the substituents of the other end of the molecule but the **C** – NH_2 carbon atom will be significantly deshielded by its proximity to so many, electronegative, oxygen atoms and this suggests that the two peaks will appear very close together.

This is what we observe so the structure of the molecule must be Candidate II.

For completeness, however we must consider the ¹H NMR spectrum.

¹H NMR Spectrum

The expanded spectrum is shown below:

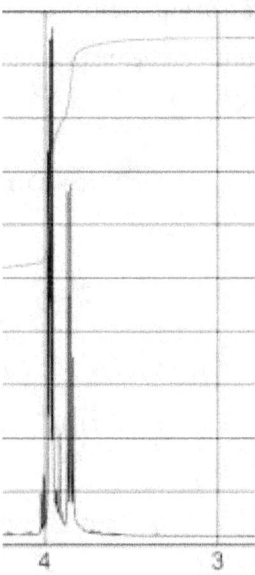

Due to the resolution (analysed at 360 MHz and in D_2O) and overlapping multiplets this spectrum is difficult to interpret but we can make the following observations:

- The broad peak at δ 4.85 ppm of integral three is due to protonated water (^1H_2O) in the solvent or impurities in the sample of the amino acid. This is a standard observation for a broad peak of variable integral in ^1H NMR spectra;

- The multiplet at δ 3.98 ppm is of integral two;

- The multiplet at δ 3.85 ppm is of integral one.

There is no signal due to the carboxylic hydrogen, $-CO_2H$ hydrogen which is not unexpected but we can now draw more conclusions.

If we expand the ^1H NMR spectrum even more we observe some detail in the multiplets:

- The multiplet at δ 3.98 ppm of integral two must be due to the $-CH_2-$ as circled below:

This multiplet is a doublet of doublets and this indicates that the two $- CH_2 -$ hydrogen atoms are *chemically and magnetically* **non-equivalent**. In other words, this means that each of these two hydrogen atoms split each other into a doublet and each of these doublets is then split into a doublet by the $-CH-$ hydrogen atom to its right as we look at the molecule as displayed above.

- The multiplet at δ 3.85 ppm, of integral one, can be assigned to the hydrogen atom on the adjacent carbon $- C(\mathbf{H}) - R_2$) is also split into a doublet of doublets by the chemically and magnetically non-equivalent hydrogen atoms on the highlighted carbon atom.

This then requires us to establish the identities of the R_1 and R_2 groups.

It is clear that one of the groups must be a hydroxyl, $-OH$, group whilst the other is an amino, $-NH_2$, group. To establish which is which we need to consider the electronegativity of the bonding substituent atoms as elctronegative atoms will deshield hydrogen atoms on the same carbon atom, increasing the observed chemical shift. Oxygen is more electronegative than nitrogen (3.5 against 3 on the Pauling electronegativity scale). This means that R_1 is $-OH$ whilst R_2 is $-NH_2$.

Conclusions

Structure:

Systematic name: 2-Amino-3-hydroxypropanoic acid

Chapter V

L - Asparagine

Isolated, in 1806, from asparagus extracts by Louis Nicolas Vauquelin and Pierre Jean Robiquet, L – asparagine was the first amino acid to have been isolated.

L – asparagine is not an essential amino acid since it is found in multiple sources such as dairy products, beef and poultry, fish and seafood as well as potatoes, nuts and many types of grain.

It took decades for the molecular structure and the determination of its structure was only established by Arnaldo Piutti in 1888.

It is also interesting since the, high temperature, reaction between L – asparagine and reducing sugars produces acrylamide which adds flavour to chips and toasted bread.

L – asparagine has a melting point of 234°C and a boiling point of 438°C. It rotates the plane of polarised light by + 5°.

With a **formula mass** (M_r) of 132.12 gmol^{-1} , this compound has the **elemental composition**: C: 36.33%, H: 6.12%, N: 21.20%, O: 36.33% which makes the empirical and molecular formulas both $C_4H_8N_2O_3$.

Infrared Spectrum

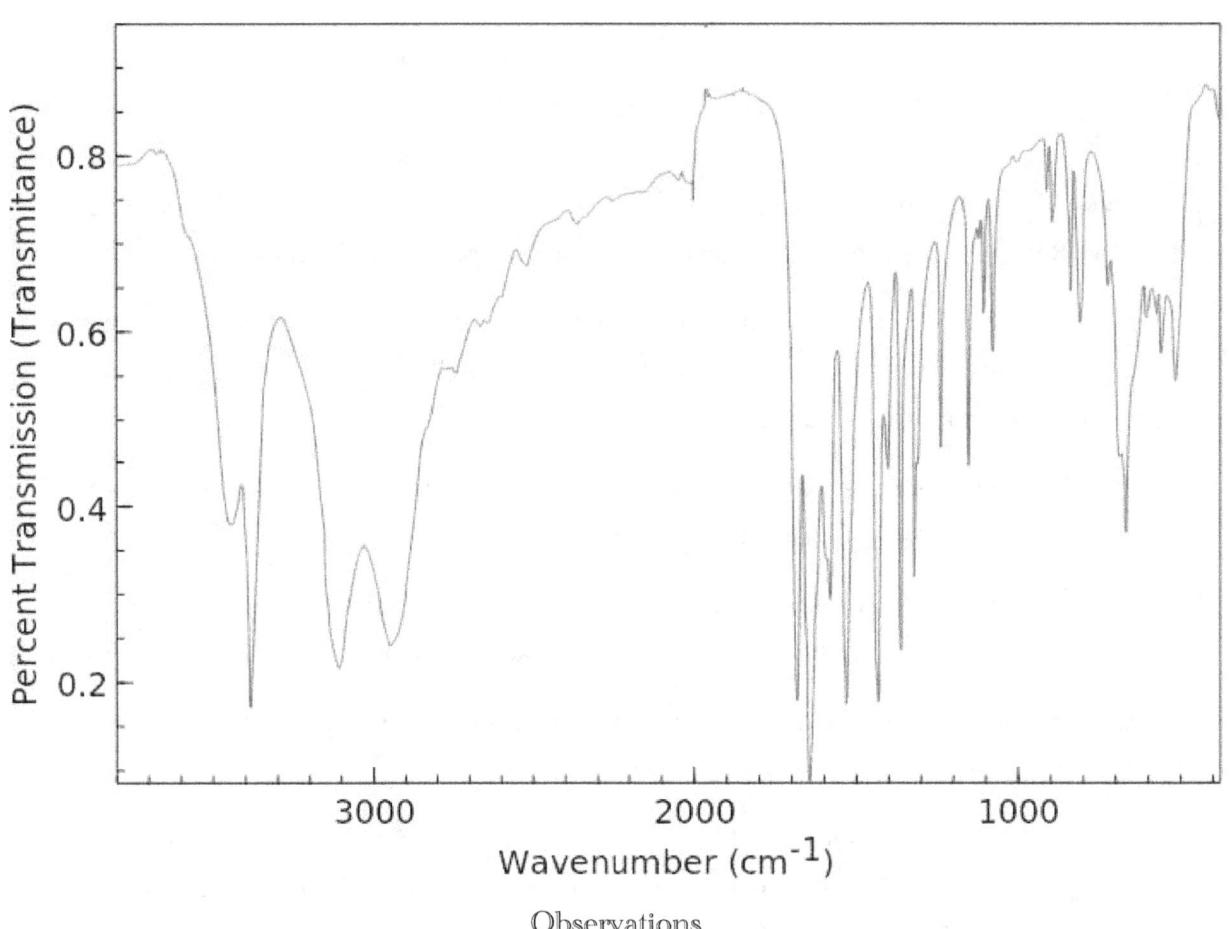

Observations

(√ / X)	Wavenumber range (cm⁻¹)	Wavenumber (cm⁻¹)	Assignment
?	3200 - 3700	3500 / 3400	**O – H**
?	3200 - 3600	3100	**N – H**
X	3000 – 3300		**C – H (aromatic)**
√	2500 – 3000	2900	**C – H (aliphatic)**
X	2200 – 2500		**C ≡ N**
?	1700 – 1800	1700	**C = O**
?	1700 – 1800	1700	**C = N**
√	1600 – 1700	1650	**C = C (aliphatic)**
X	1585 – 1600		**C – C (aromatic)**
X	1450 – 1600		**C – C (aromatic)**
X	1000 – 1300		**C – O**
X	700 – 1000		**C – X (X = Cl, Br or I)**

Conclusions

From this spectrum,

- It appears that the molecule contains either or both N – H and O – H groups.

- That there is a stretch in the aromatic C – H region but the molecule is too small to be aromatic.

- There could be C = N, C = O and C = C groups. That appears unlikely, again due to the size of the molecule and ww can learn more from analysing the mass spectrum which is our next task.

L - Asparagine

Mass Spectrum

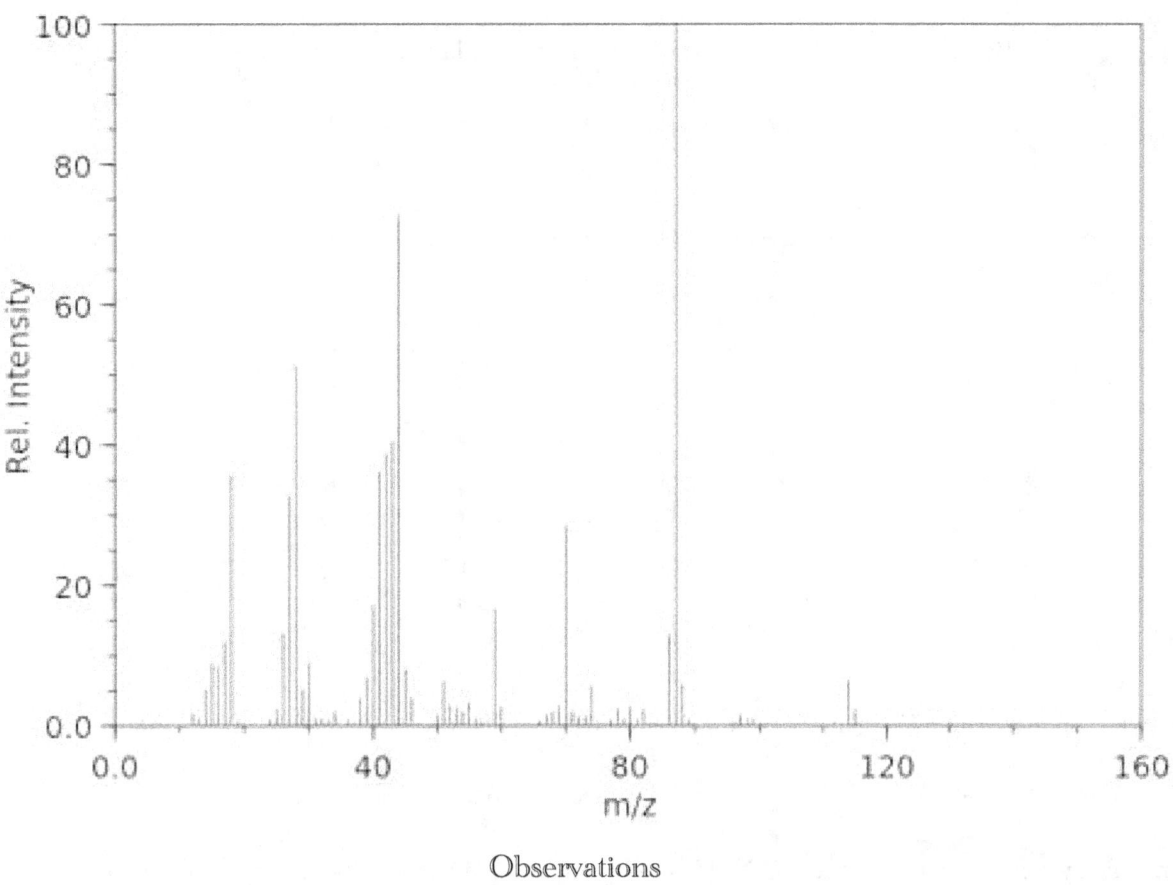

Observations

Charged fragments (m/z)	Assignment	Charged fragments (m/z)	Assignment
Molecular ion: 114	$[C_4H_6N_2O_2]^+$	Base peak: 87	$[C_2H_3N_2O_2]^+$

73	$[C_2H_4NO_2]^+$	51	$[C_4H_3]^+$
70	$[C_4H_6O]^+$	44	$[C_2H_4O]^+$
59	$[C_3H_7O]^+$	28	$[C_2H_4]^+$

Conclusions

The mass spectrum is complicated and there are a lot of peaks which cannot be assigned. It is clear, however, that molecule has a carbon chain and the peak at m/z = 70 indicates the presence of a carbon chain with a carboxylic acid group. This means that, as a first guess, we can tentatively assign the molecule to have the following basic structure:

but we can deduce the structure from the ^{1}H and ^{13}C NMR spectra.

NMR Spectra

The ^1H and ^{13}C NMR spectra, measured in ^2H_2O (D$_2$O) are displayed below:

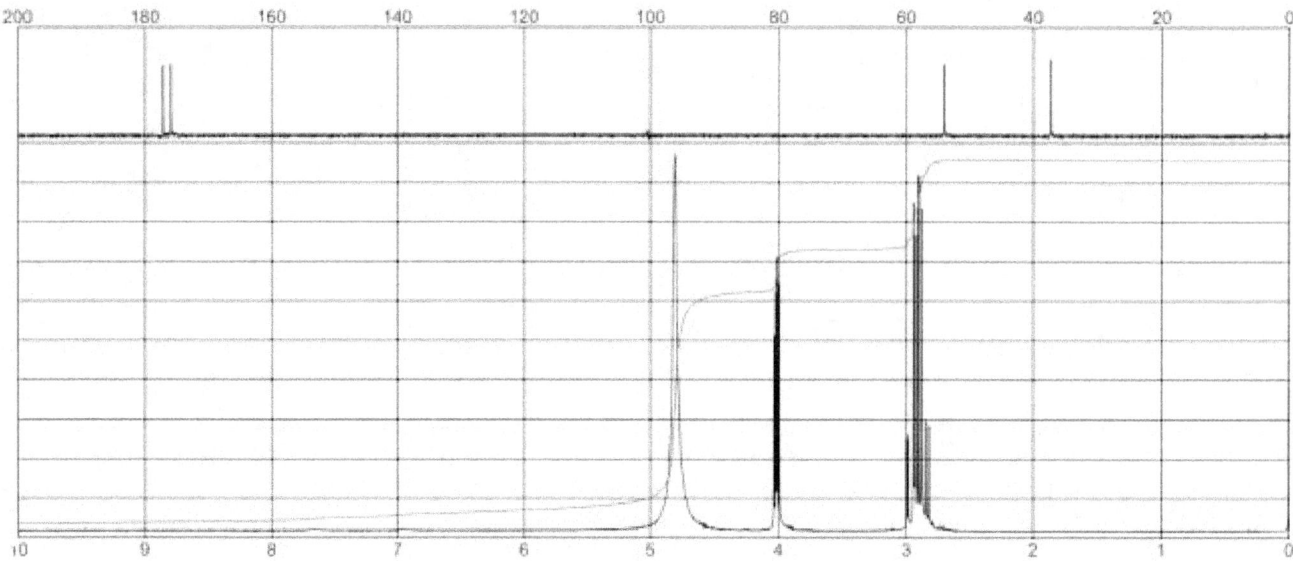

and we will consider the ^{13}C NMR spectrum first as it contains only four peaks in very distinct regions.

^{13}C NMR Spectrum

We can readily analyse part of the ^{13}C NMR spectrum but the two peaks of lower chemical shift are harder to assign. In summary;

* There are four singlets all of integral one;
* Two of the singlets are in the δ 160 – 220 ppm region and indicate that the molecule contains two C = O functional groups;
* The peak at δ 54 ppm is either due to a C – O or a C – N bond;
* The peak at δ 38 ppm can be assigned to either a C – N or a C – C group.

Since we have deduced from the mass spectrum that the molecule has a four carbon chain and two C = O functional groups the immediate tentative assumption could be that the molecule is of the following form (where the squiggles represent groups yet to be determined):

This structure accounts for the presence of one or more hydroxyl, – OH, groups whose presence was indicated by the infrared spectrum and this could account for the peak at m/z = 114 other than for the presence of too many oxygen atoms.

This means that we must remove one of the – OH groups and so the question then is what to put in its place.

A reasonable proposal would be to replace it with a nitrogen atom which would then give us the following basic structure:

This seems sensible as the infra red spectrum indicates the presence of one or more N – H groups and there are two nitrogen atoms in the molecular formula.

This accounts for the following partial formula, $C_4H_3NO_3$, leaving H_5N to fit in somewhere where the squiggles are drawn. It is clear that the nitrogen atom must be part of an amino, $-NH_2$, group.

We, therefore, have two candidate structures as shown below:

I

II

Both these structures will produce very similar ^{13}C NMR spectra and the fourth ^{13}C NMR signal (δ 38 ppm) can be accounted for by the C – C bond which is present in both structures. This means that we must reconsider the other spectral and spectrometric data especially the mass spectrum.

The mass spectrum demonstrates a base peak of m/z = 87.

This can only be accounted for by the fragmentation as shown highlighted where the numbers are the relative atomic mass or formula masses of the substituents in the fragment:

No other fragment matches the base peak so the molecular structure must be:

We must, however, also consider the ^1H NMR spectrum to confirm this structure and that is our final task.

^{1}H NMR Spectrum

We can concentrate on the δ 2 – 4.5 ppm region and observe the following:

▪ The multiplet at δ 4.02 ppm, of integral one, would appear, at low resolution, to be a quartet but close inspection reveals that it is actually a doublet of doublets;

▪ The apparent octet, of integral two, centred on δ 2.9 ppm actually comprises a

　▪ Doublet of doublets - the central four peaks – of integral one;

　▪ Doublet of doublets where the doublets are the pairs of peaks on either side of the multiplet also of integral one.

We can explain these multiplets as shown below where the relevant hydrogen atoms are labelled alphabetically.

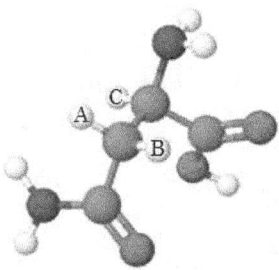

This becomes even clearer if we examine a three-dimensional model as shown below. For clarity H_a, H_b and H_c are labelled A, B and C respectively:

▪ It becomes immediately evident that H_a and H_b (labelled A and B) cannot possibly be chemically and magnetically equivalent since there is a planar –C=O(NH_2) group. Since the C=O bond cannot rotate, H_a and H_b immediately become chemically and magnetically *non-equivalent*.

▪ At low resolution H_a/H_b would produce a doublet of integral two due to splitting by H_c but at higher resolution we observe that H_a is split into a doublet by H_b which is then split into a doublet by H_c. Similarly, H_b will be split by H_a and then by H_c. We can also describe it as splitting by H_c and then by the other of the pair H_a/H_b. It makes no difference as the conclusion is the same.

Conclusions

Structure:

Systematic name: L – asparagine and 2-amino-3-carbamoylpropanoic acid (IUPAC).

Chapter VI

L – Valine

L – valine was first isolated from casein in **1901** by Hermann Emil Fischer although the name comes from valeric acid (pentanoic acid)which was first extracted from the roots of the plant valerian.

L – valine is synthesised by plants but not by animals. It is, therefore, classified as an essential amino acid in animals, and must be incorporated into the diet. It is found in protein-containing foods such as meat, dairy products, beans and potatoes.

It is also interesting that the, multi-step, synthesis in plants starts from pyruvic acid and this process also produces L – leucine (Chapter VII).

L – valine is essential for the building of proteins but is also associated with weight loss and diabetic problems.

L – valine has a melting point of 298°C when it decomposes and demonstrates an angle of rotation of plane polarised light of +23°.

With a **formula mass** (M_r) of 117.15 g mol^{-1} , this compound has the **elemental composition**: C: 51.22 %, H: 9.48 %, N: 11.96 %, O: 27.32 % meaning that both the empirical and molecular formulas are $C_5H_{11}NO_2$

L - Valine

Infrared Spectrum

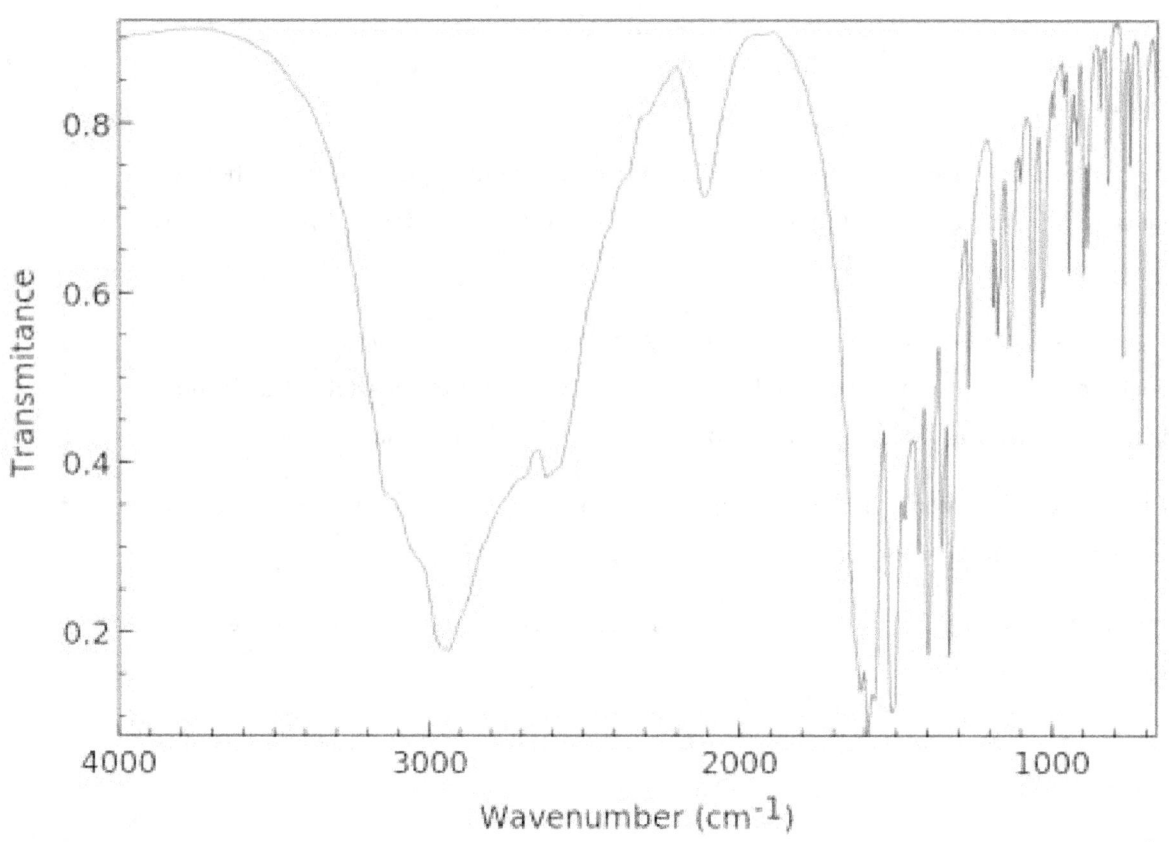

Observations

(√ / X)	Wavenumber range (cm⁻¹)	Wavenumber (cm⁻¹)	Assignment
?	3200 - 3700	3000 – 3200	O – H
?	3200 - 3600	3000 – 3200	N – H
X	3000 – 3300		C – H (aromatic)
√	2500 – 3000	2500 – 2950	C – H (aliphatic)
X	2200 – 2500		C ≡ N
X	1700 – 1800		C = O
X	1700 – 1800		C = N
√	1600 – 1700	1600	C = C (aliphatic)
X	1585 – 1600		C – C (aromatic)
X	1450 – 1600		C – C (aromatic)
√	1000 – 1300	Fingerprint region	C – O
X	700 – 1000		C – X (X = Cl, Br or I)

Conclusions

This is a very difficult spectrum to analyse as the stretch centred on 2950 cm⁻¹ is very broad.

Given that the molecule contains both nitrogen and oxygen atoms experience should tell that it is highly possible for O – H and N – H groups to exist.

The spectrum also implies the presence of a C = C bond but the presence of two oxygen atoms implies that a carboxylic acid, –CO₂H, might be present. We can learn more from the mass and NMR spectra.

Mass Spectrum

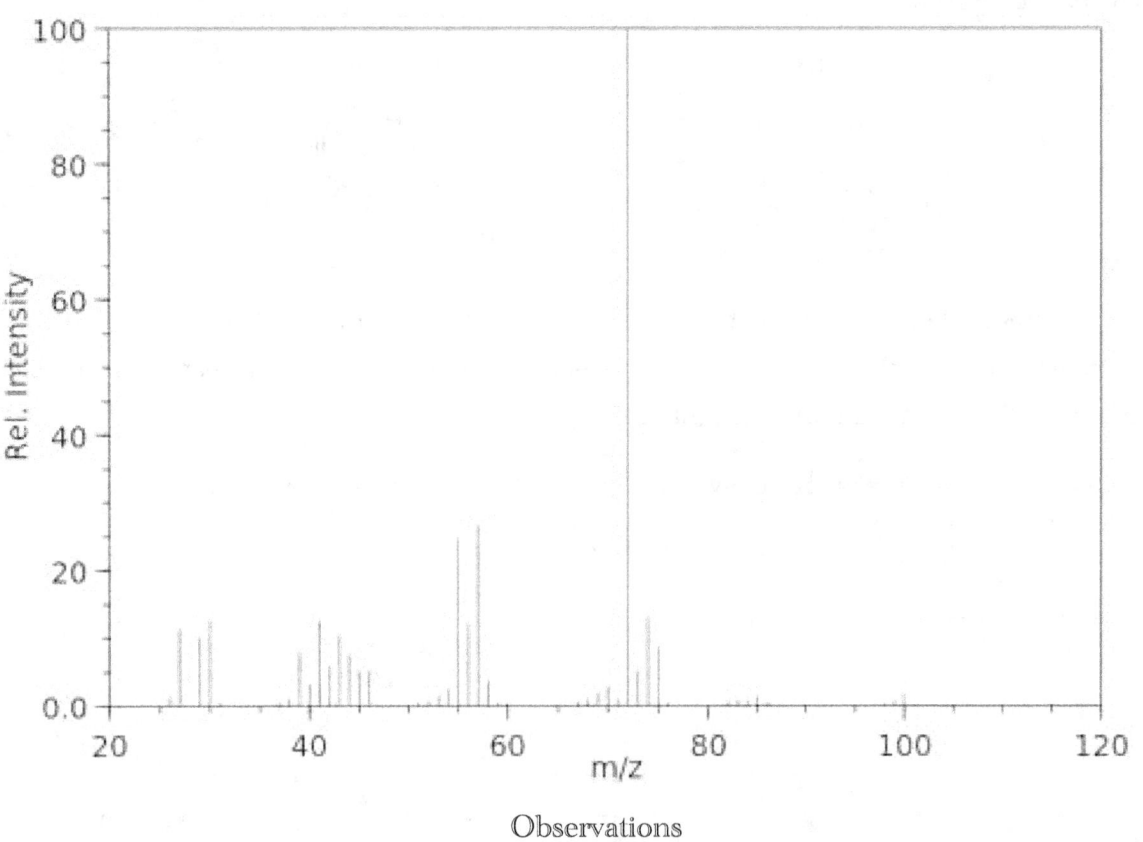

Observations

Charged fragments (m/z)	Assignment	Charged fragments (m/z)	Assignment
Molecular ion: 117	$[C_5H_{11}NO_2]^+$	**Base peak: 72**	$[C_4H_8O]^+$

74	$[C_3H_6O_2]^+$	55	$[C_4H_7]^+$
70	$[C_5H_{10}]^+$	43	$[C_3H_7]^+$ or $[C_2H_3O]^+$
57	$[C_4H_9]^+$ or $[C_3H_5O]^+$	29	$[C_2H_5]^+$

Conclusions

This is a very interesting spectrum and the presence of the base peak at m/z = 72 indicates either a four – carbon chain or a substituted three – carbon chain.

The peak at m/z = 74 shows a three carbon chain with two oxygen atoms suggesting that the molecule contains a carboxylic acid, $-CO_2H$, group.

We can, tentatively, make a start on the structure of the molecule with an educated guess which is often the best way. There are two possibilities for the basic skeletal formula:

In both cases we have accounted for $C_4H_6O_2$ from a formula of $C_5H_{11}NO_2$ leaving us with CH_5N to account for.

These atoms can be assigned by writing the atoms as two groups: CH₃ and NH₂ and we then have two possible isomers, now numbered III and IV:

III   **IV**

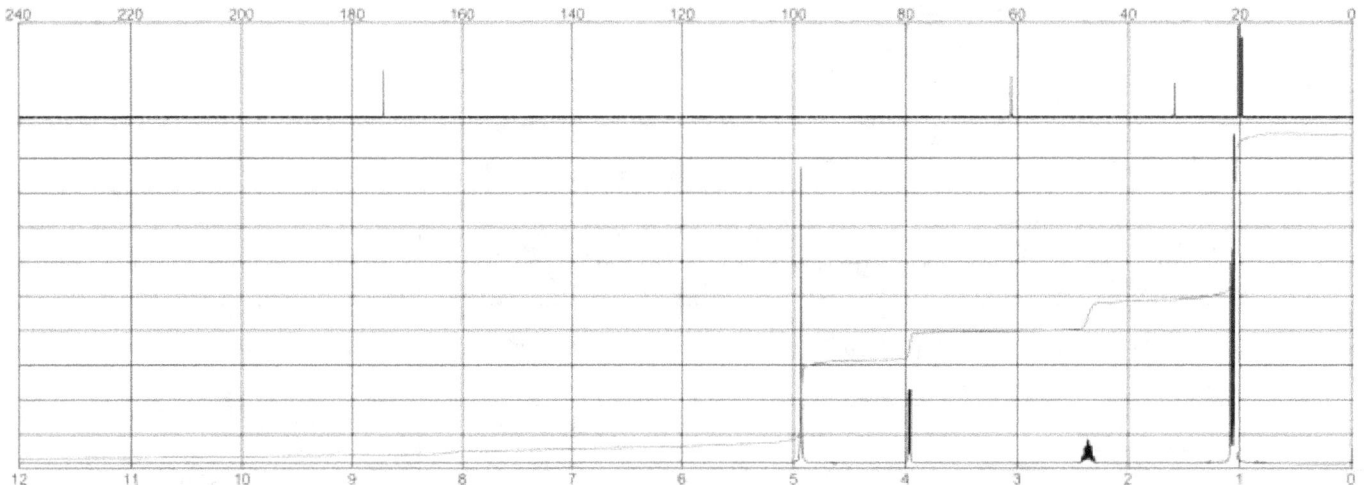

Candidate IV is impossible as it contains two chiral centres (indicated by * and remember that there is a hydrogen atom attached to each of the two carbon atoms which is not shown). This leaves us with candidate III but we must also bear in mind that we might be completely wrong.

We can establish if the structure is correct by examining the ^{1}H and ^{13}C NMR spectra.

NMR Spectra

The ^{1}H and ^{13}C NMR spectra are displayed below:

It is better, however, to predict the spectrum first and we start by predicting the ^{13}C NMR spectrum first.

^{13}C NMR Spectrum

Looking at the candidate structure, repeated below but now labelled alphabetically,

we can predict that we will observe:

- A singlet of integral two due to C_a and $C_{a'}$ (which are chemically and magnetically equivalent) peak in the δ 0 – 40 ppm region;
- A singlet of integral one, also in the δ 0 – 40 ppm region assignable to C_b;
- A singlet of integral one in the range δ 30 – 70 ppm due to Cc and finally;
- A singlet of integral one in the δ 160 – 220 ppm due to C_d.

This is exactly what we observe as tabled below:

Carbon atom	Chemical shift δ (ppm)
C_a / $C_{a'}$	19
C_b	31
C_c	62
C_d	175

but we must also confirm the structure by examining the ^1H nmr spectrum and we will predict it first.

^1H NMR Spectrum

We will use the following structure which is now labelled for the hydrogen atoms.

There is lots of coupling in this structure and we can predict the following:

- H_e and $H_{e'}$ will produce a signal of integral six. H_e is split by H_f and then by the three chemically and magnetically equivalent $H_{e'}$ producing a triplet. The same applies to $H_{e'}$.

- H_f will produce a peak of integral one and will be a quartet due to splitting by H_e which will then be split into an octet by $H_{e'}$ and this octet will then be split by H_g producing a complex multiplet which is shown below in greatly expanded form:

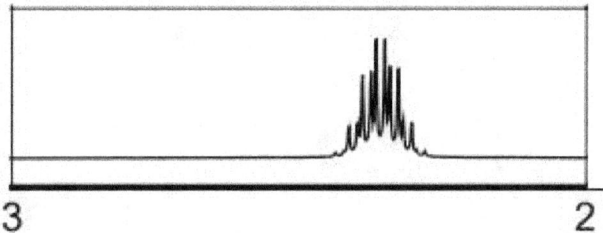

- H_g will produce a doublet of integral one due to splitting by H_f.

- H_h will produce a singlet of integral two anywhere in the δ 0 – 12 ppm.

- H_i, being a hydroxyl hydrogen, may appear anywhere between δ 0 and δ 12 ppm but may not appear at all. If it does then it will have an integral of one.

Hydrogen atom(s)	Chemical shift (ppm)	Multiplicity	Integral
H_e / $H_{e'}$	1.10	Triplet	6
H_f	2.45	Complex multiplet	1
H_g	3.95	Doublet	1
H_h	4.95	Singlet	2
H_i	Absent		

This completely matches the prediction and, as is common, the hydroxyl hydrogen did not appear.

L – Valine

Conclusions

Structure:

Systematic name: L - valine and also 2-amino-3-methylbutanoic acid (IUPAC).

Chapter VII

L – Leucine

Discovered in 1819 by Joseph-Louis Proust as half of a racemic mixture, L – leucine is an essential amino acid that is used by the body to produce protein and repair muscle damage.

An essential amino acid, as the human body cannot produce it, L – leucine is found in all protein containing foods such as meat, dairy products, soy products and beans and is also used as a food additive and as a flavour enhancer (E641).

Medically, both L – leucine and its mirror image, D – leucine, have been found to be useful in preventing seizures in mice but other health claims have not been proven.

L – leucine has a melting point of 286°C, at which point it decomposes, and rotates the plane of polarised light by -11°C.

With a **formula mass** (M_r) of 131.18 g mol^{-1}, L – leucine has the **elemental composition**: C: 54.89%, H: 10.10 %, N: 10.68%, O: 24.39 % and so the empirical and molecular formulas are both $C_6H_{13}NO_2$.

Infrared Spectrum

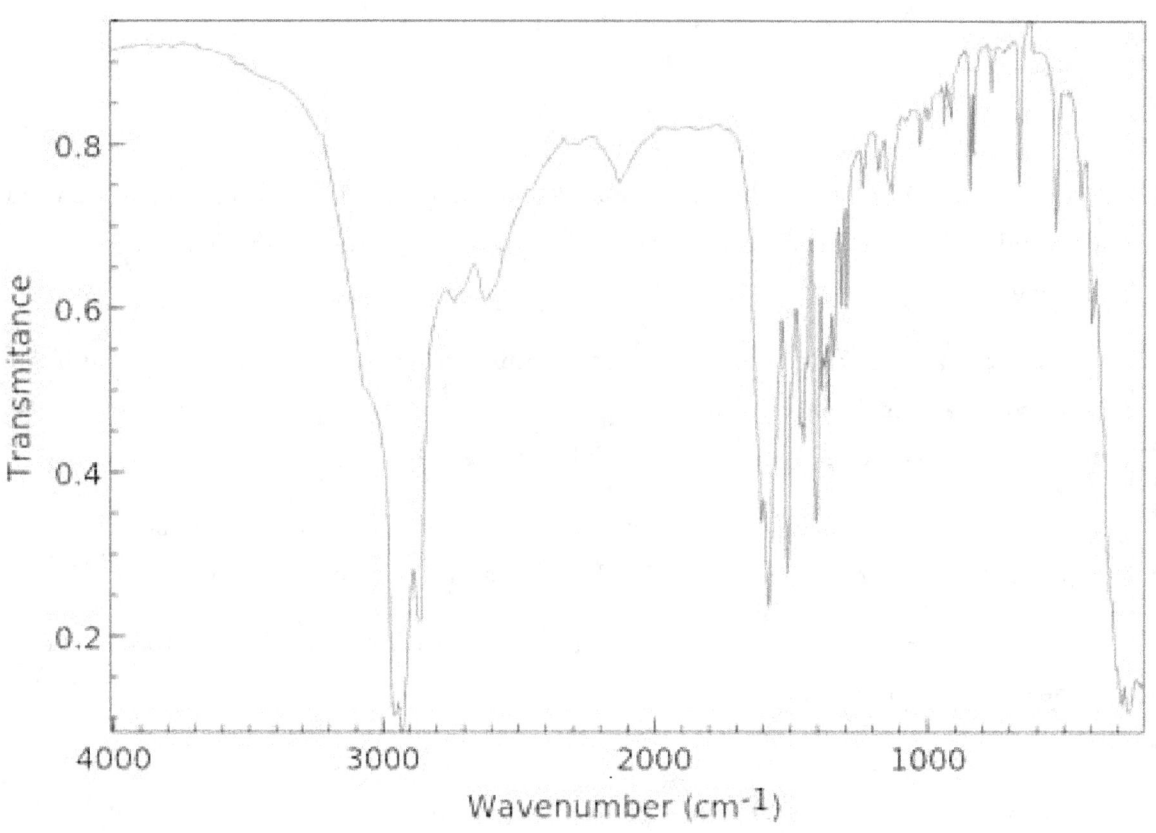

Observations

(√ / X)	Wavenumber range (cm⁻¹)	Wavenumber (cm⁻¹)	Assignment
X	3200 - 3700		O – H
X	3200 - 3600		N – H
X	3000 – 3300		C – H (aromatic)
√	2500 – 3000	2980, 2920	C – H (aliphatic)
X	2200 – 2500		C ≡ N
X	1700 – 1800		C = O
X	1700 – 1800		C = N
√	1600 – 1700	1620, 1580	C = C (aliphatic)
X	1585 – 1600		C – C (aromatic)
X	1450 – 1600		C – C (aromatic)
X	1000 – 1300	Fingerprint region	C – O
X	700 – 1000		C – **X** (X = Cl, Br or I)

Conclusions

This molecule is clearly aliphatic (from the peaks below i.e. to the right of 3000 cm⁻¹ and it also appears that it contains a C=C bond however there are indications from the fingerprint region that there is a C – O bond which, from the formula implies the presence of a carboxylic acid, – CO_2H, group.

We can learn much more from the mass spectrum and the ¹H and ¹³C NMR spectra.

L - Leucine

Mass Spectrum

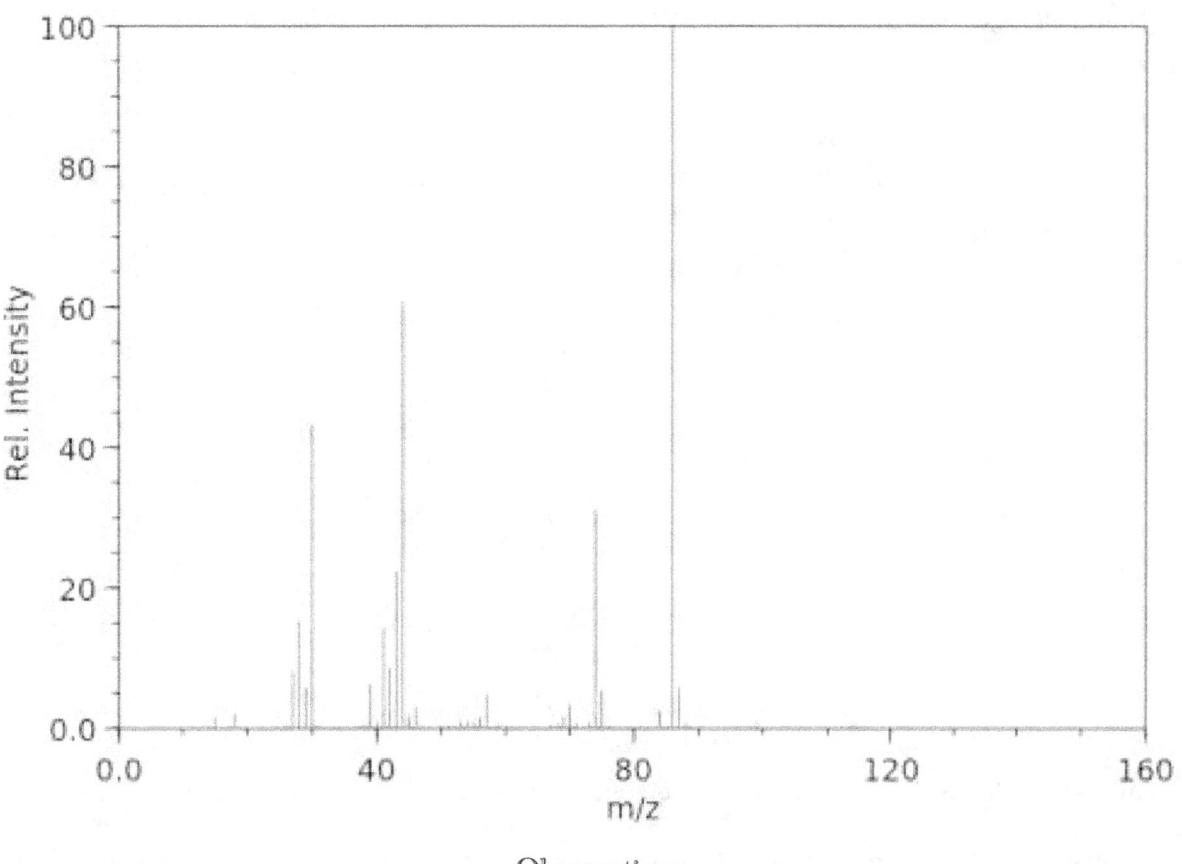

Observations

Charged fragments (m/z)	Assignment	Charged fragments (m/z)	Assignment
Molecular ion: 86	$[C_5H_{10}O]^+$	Base peak: 86	$[C_5H_{10}O]^+$

74	$[C_4H_{10}O]^+$	44	$[CH_3CHNH_2]^+$
57	$[C_4H_9]^+ / [C_2H_5C=O]^+$	30	$[CH_2NH_2]^+$

Conclusions

The molecular ion which is the same as the base peak tells us little however:

- The peak at m/z=74 indicates a chain of at least four carbon atoms;
- As does the peak at m/z=57 which, however, could also be due to a carboxylic acid, $-CO_2H$ whilst;
- The peaks at m/z = 44 and 30 indicate the presence of a terminal $-NH_2$ group.

This gives us a range of possibilities for the structure, the simplest being

This, however, cannot be correct as all but the simplest amino acids are chiral and this structure is achiral. The carboxylic acid group must be terminal and this means that the amino group, $-NH_2$, must be a substituent on the chain but the question is where? Only the ^1H and ^{13}C NMR spectra can tell us that.

NMR Spectra

The ^{1}H and ^{13}C NMR spectra are displayed below:

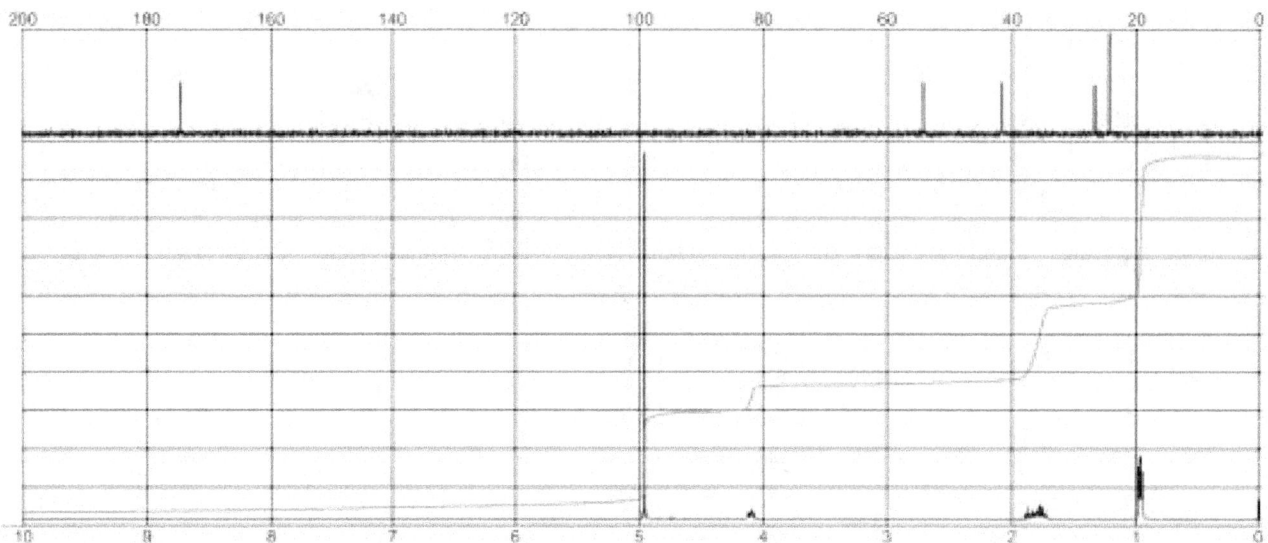

Observations

^{13}C NMR Spectrum

We will start with the ^{13}C NMR spectrum from which we can make the following immediate observations:

- The peak of integral one at δ 175 ppm indicates the presence of a C=O bond which contradicts the ir spectrum;
- The peak of integral one at δ 54 ppm is indicative of a C – N group whilst;
- That at δ 42 ppm represents a C – C bond and;
- The peaks of integral one and two, respectively, at δ 28 and δ 25 ppm indicate the presence of C – C linkage but what is more important is the peak of integral two which suggests the presence of two methyl groups which must be terminal.

Naturally, this leads us to the following two possible structures:

The molecule must be chiral, as all but the simplest amino acid (glycine – Chapter III), are all (L –) chiral molecules so these are the only possibilities.

We must examine the ^{1}H NMR spectrum to determine which structure is correct.

L - Leucine

^{1}H NMR Spectrum

An expanded view of the spectra (δ 0 – 5 ppm) is shown below.

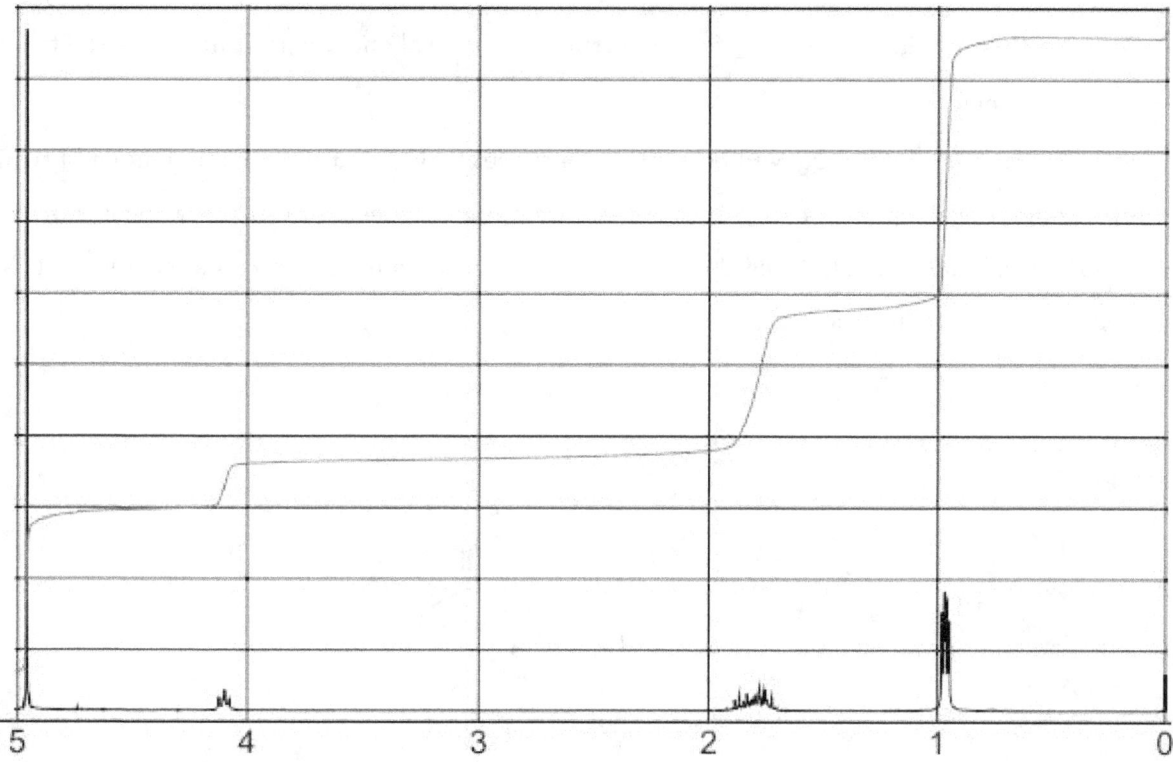

■ The doublet of quartets at δ 0.95 ppm, of integral six, is due to alkyl hydrogen atoms i.e. $-CH_3$. Both candidate molecules will produce this signal since both have six, terminal, methyl hydrogen atoms and these group $-CH_3$ groups are bonded to a carbon atom with one hydrogen atom bonded to it. This is interesting since it demonstrates that whilst the three hydrogen atoms within the individual alkyl groups are chemically and magnetically equivalent the two alkyl groups are not chemically and magnetically equivalent.

For example, for candidate I, H_a will be split into a quartet by H_b and the quartet will be split by H_c. Similarly, H_b will be split into a quartet by H_a and this quartet will be split into a doublet of quartets by H_c. The same rationale applies, in candidate II to H_h, H_i and H_j. Whilst this signal does not help us to distinguish between the two candidate molecules it does show that we are on the right track.

■ The complex multiplet at δ 1.9 ppm of integral three can be assigned to H_c which is split by H_a, H_b and H_d. There is no cause of a complex multiplet of integral three in candidate II. H_k will produce a doublet of triplets or a triplet of doublets due to splitting by H_j and H_m.

This is the first evidence that candidate I is the correct structure.

⌗ What really does help us to identify the correct structure is the signal due to the hydrogen bonded to the carbon which is also bonded to the amino, –NH$_2$, group. This is H$_e$ in candidate I and H$_k$ in candidate II.

> ⌗ In candidate I, H$_e$ will produce a triplet of integral one due to splitting by the two hydrogen atoms, H$_d$.

> ⌗ In candidate II, H$_k$ will be split into a triplet by H$_m$ and then into a doublet of triplets by H$_j$. Although the signal at δ 4.15 ppm is outside the range suggested in the data sheet this must be due to H$_e$ (candidate I) or H$_k$ (candidate II). This signal is a triplet and so must be due to H$_e$ and so the correct structure is candidate I.

Conclusions

Structure:

Systematic name: L – leucine and 2–amino–4–methylpentanoic acid.

Chapter VIII

L – Glutamine

Isolated from sugar beet as a racemic mixture in 1883 by Ernst Schulze and Emil Bosshard, L – glutamine is the most abundant of the so called, naturally occurring, non-essential amino acids and is one of the few amino acids that can cross the blood – brain barrier.

A non-essential amino acid since the body produces L – glutamine through catabolism* of protein-containing foods, it is essential for protein growth in babies and has been found to have some efficacy in mitigating the symptoms of sickle cell disease.

L – glutamine is found protein-rich foods such as meat, dairy products as well as in green vegetables, carrots and parsley.

It has been found to play a role in many biochemical processes such as protein and lipid synthesis, as a source of cellular energy and nitrogen donation as well as in the citric acid cycle. It is mainly used in intestinal and kidney cells but has also been shown to be needed by cancer cells.

L – glutamine has a melting point of 185°C when it decomposes and rotates the angle of plane polarised light by + 6.5°.

With a **formula mass** (M_r) of 146.15 gmol^{-1} , this compound has the **elemental composition**: C: 41.05%, H: 6.91%, N: 19.17%, O: 32.84% and both the empirical and molecular formulas are $C_5H_{10}N_2O_3$.

*Catabolism** is the process by which large molecules are broken down, enzymatically, to produce smaller molecules or fragments and biochemical energy which is conserved in molecules such as adenosine triphosphate (ATP). This process is part of the Krebs Cycle.

L – Glutamine

Infrared Spectrum

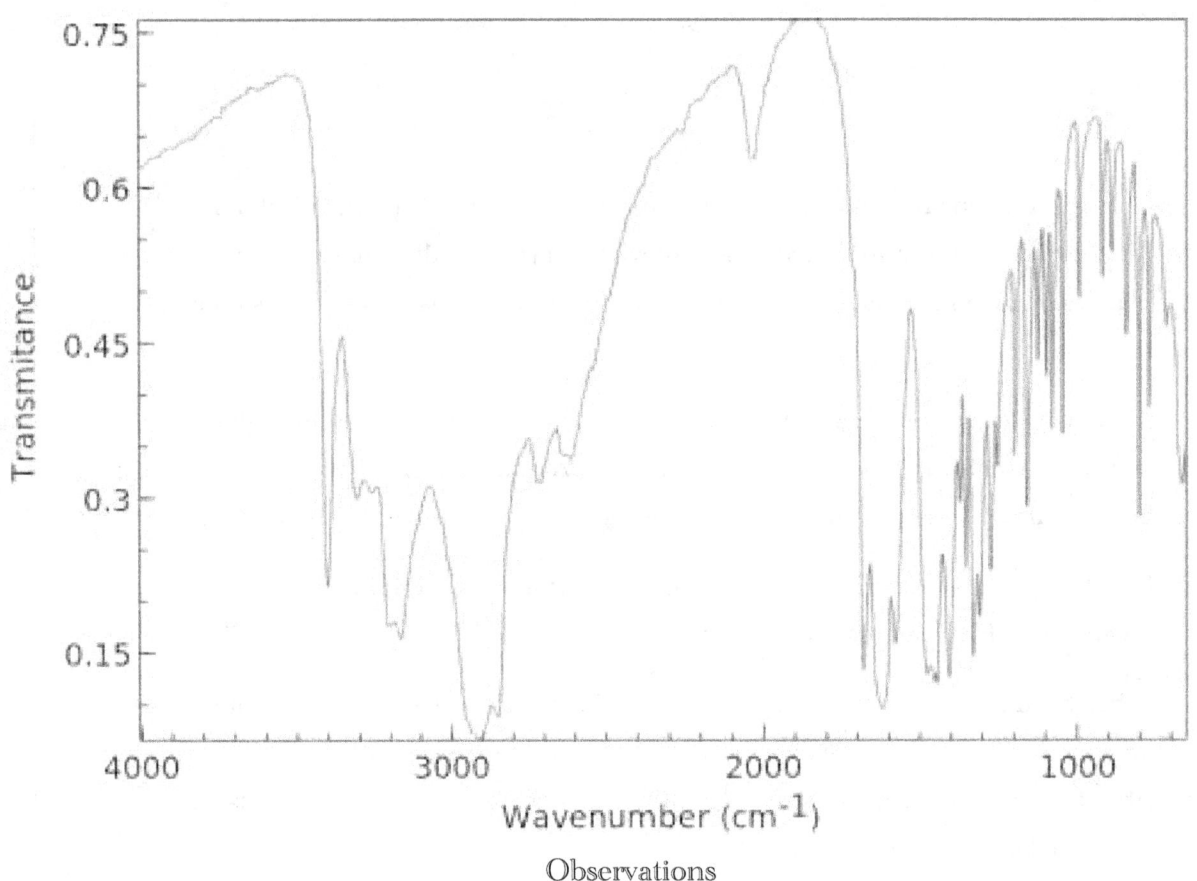

Observations

(√ / X)	Wavenumber range (cm⁻¹)	Wavenumber (cm⁻¹)	Assignment
?	3200 - 3700	3450, 3300, 3200, 3180	O – H
?	3200 - 3600	3450, 3300, 3200, 3180	N – H
X	3000 – 3300		C – H (aromatic)
√	2500 – 3000	2900	C – H (aliphatic)
X	2200 – 2500		C ≡ N
?	1700 – 1800	1700	C = O
?	1700 – 1800	1700	C = N
√	1600 – 1700	1620	C = C (aliphatic)
X	1585 – 1600		C – C (aromatic)
X	1450 – 1600		C – C (aromatic)
√	1000 – 1300	Fingerprint region	C – O
X	700 – 1000		C – X (X = Cl, Br or I)

Conclusions

- There are peaks in the aromatic C – H region but, given the molecular formula of the compound, ($C_5H_{10}N_2O_3$), it is highly unlikely that the compound is aromatic.

- There is a peak at 3450 cm⁻¹ that could be due to either an O – H or a N – H bond or both.

- Given that the molecule contains three oxygen atoms then it is highly likely that the molecule contains a carboxylic acid, –CO_2H, group. This would also produce a peak due to a C – O group in the fingerprint region (1000 – 1300 cm⁻¹) but it is impossible to assign a particular peak to this bond.

Mass Spectrum

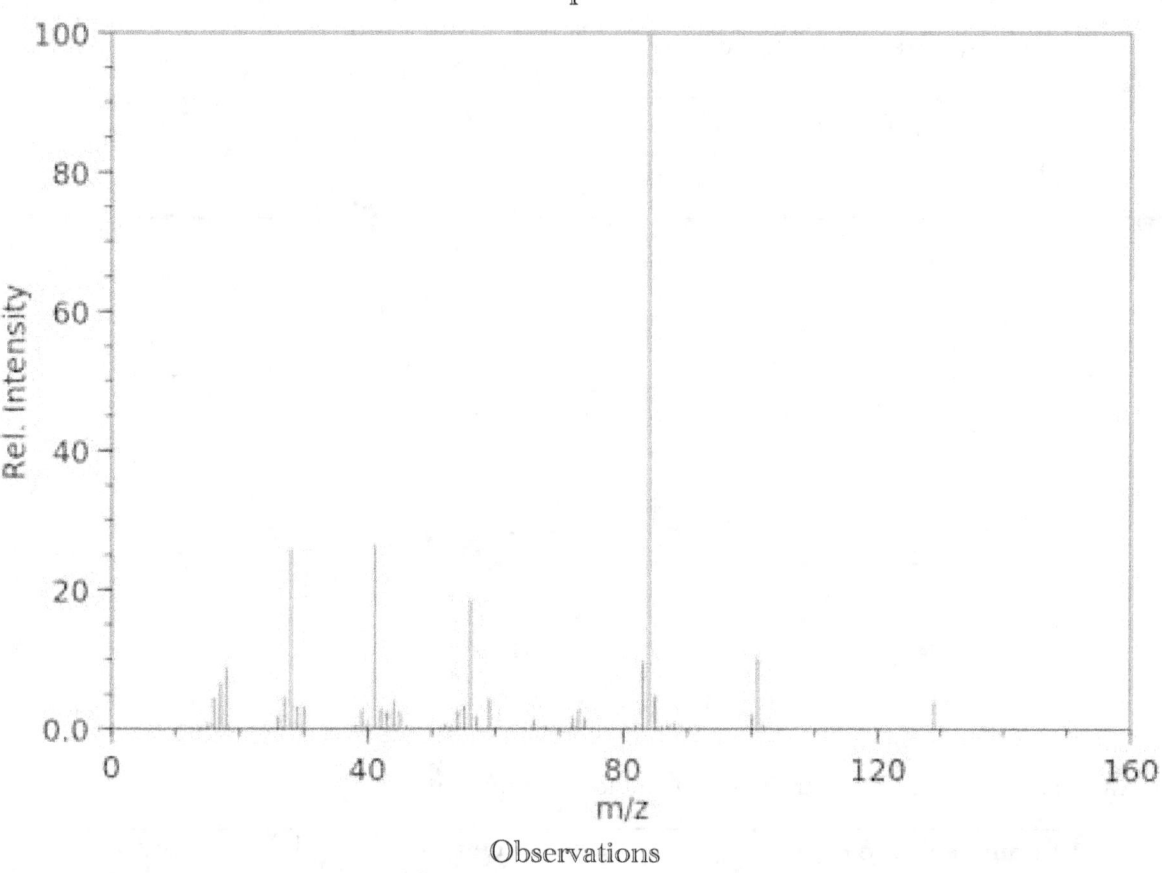

Observations

Charged fragments (m/z)	Assignment	Charged fragments (m/z)	Assignment
Molecular ion: 129	$[C_5H_9N_2O_2]^+$	Base peak: 84	$[C_5H_8O]^+$

101	$[C_5H_9O_2]^+$	28	$[C_2H_4]^+$
55	$[C_4H_7]^+$	18	$[H_2O]^+$

Conclusions

The presence of both C_4 and C_5 containing fragments leads to the obvious suggestion that the molecule contains a five – carbon chain. We have already estimated that the molecule contains a carboxylic acid group which must be terminal and, given that the molecule contains two nitrogen atoms it is also reasonable to suggest that is contains an amino group and the obvious place to put that is at the other end of the molecule. This means that we have a base structure such as displayed below where the squiggles represent undetermined functional groups.

There are, of course, other possibilities such as a shorter chain, more highly substituted chain but we have to start somewhere and we can establish the correct structure from the ^1H and ^{13}C NMR spectra which is our next task.

NMR Spectra

The ^{1}H and ^{13}C NMR spectra are displayed below:

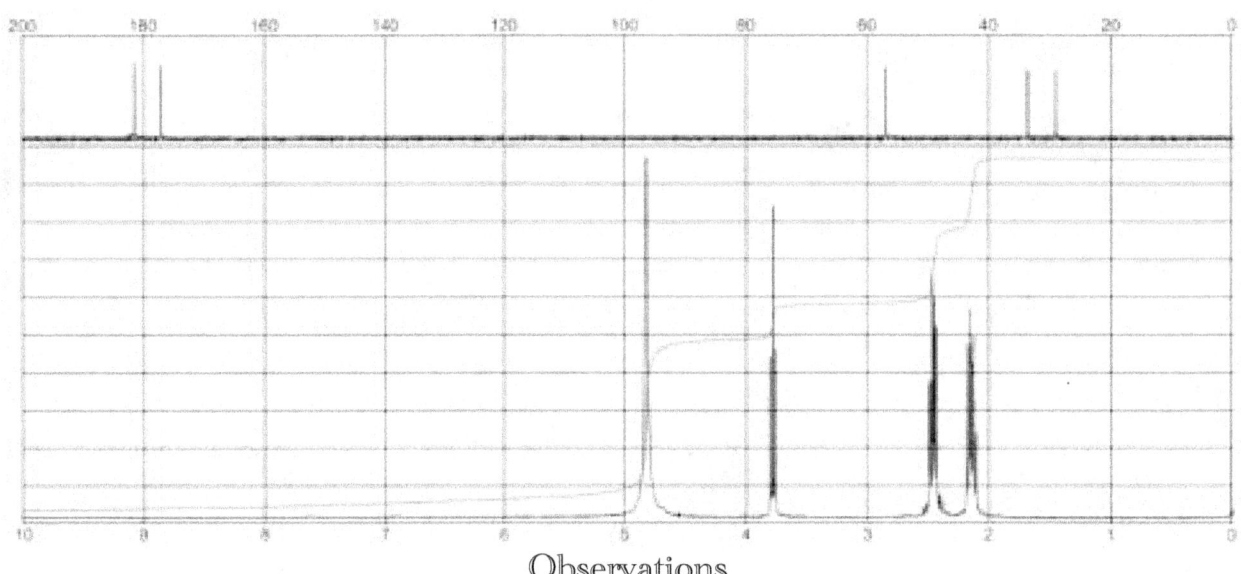

Observations

^{13}C NMR Spectrum

We will examine the ^{13}C NMR spectra and the singlets are tabulated below:

Chemical shift δ (ppm)	Assignment(s)	Integral
181	C = O or C = N	1
176	C = O or C = N	1
56	C – O	1
34	C – N	1
29	C – C	1

At first glance this spectrum appears simple to interpret but as soon as we start it becomes significantly more challenging than it appears. This is because:-

▪ An immediate assumption would be that, due to the peaks at δ 181 and δ 176 ppm, the molecule must contain two carboxylic acid, –CO_2H, groups. There is, however, only one C – O bond (δ 56 ppm). This means that there can only be one –CO_2H group.

▪ The molecule contains two nitrogen atoms and so a reasonable assumption would be that it possesses one C = N and one amino, –NH_2, groups. This accounts for three of the signals and a fourth, at d 56 ppm indicates the presence of a carboxylic acid group.

We can overcome these issues by guessing, and at this stage, it is just a guess that the molecule contains one terminal –CO_2H group and a terminal amide, – $C(O)NH_2$, group and this then gives us the following structure, where again the squiggles represent undetermined substituents:

L – Glutamine

This uses up $C_5H_3NO_2$ from the molecular formula leaving us with H_7N to place in the structure.

- Due to the signal at δ 34 ppm, there must be an amino group which uses up NH_2 leaving us with five hydrogen atoms to accommodate. One of these is easy to assign as the amino group must be a substituent on a carbon in the chain which will also have a hydrogen atom bonded to it.
- That leaves us with four hydrogen atoms which must be bonded to the two carbon atoms in the chain.

This gives us three possible candidate structures:

We can distinguish them by predicting the splitting in the ^1H nmr spectrum and then comparing our predictions with the actual spectrum.

^1H NMR Spectrum

It is easiest for us to draw the three structures again but with the relevant hydrogen atoms in the chain labelled alphabetically as shown below:

For:-

- **Candidate I:**
  - H_a will produce a triplet, of integral two, due to splitting by the two H_b hydrogen atoms;
  - H_b will produce a doublet of triplets of integral two, due to splitting by H_a and then by H_c;
  - H_c will produce a triplet of integral one due to splitting by the two H_b hydrogen atoms.
- **Candidate II:**
  - H_d will produce a doublet, of integral one, due to splitting by H_e;
  - H_e will produce a triplet of triplets, of integral one, due to splitting by H_d and then by H_f;
    We could equally predict a triplet of triplets due to splitting by H_f and then by H_d.
  - H_f will produce a doublet, of integral two, due to splitting by H_e.
- **Candidate III:**
  - H_g will produce a triplet of integral one, due to splitting by the two H_h hydrogen atoms;
  - H_h will produce a triplet of triplets, of integral two, due to splitting by H_g and H_i;
  - H_i will produce a triplet, of integral two, due to splitting by H_h.

We need to examine the expanded ¹H NMR spectrum which is shown below:

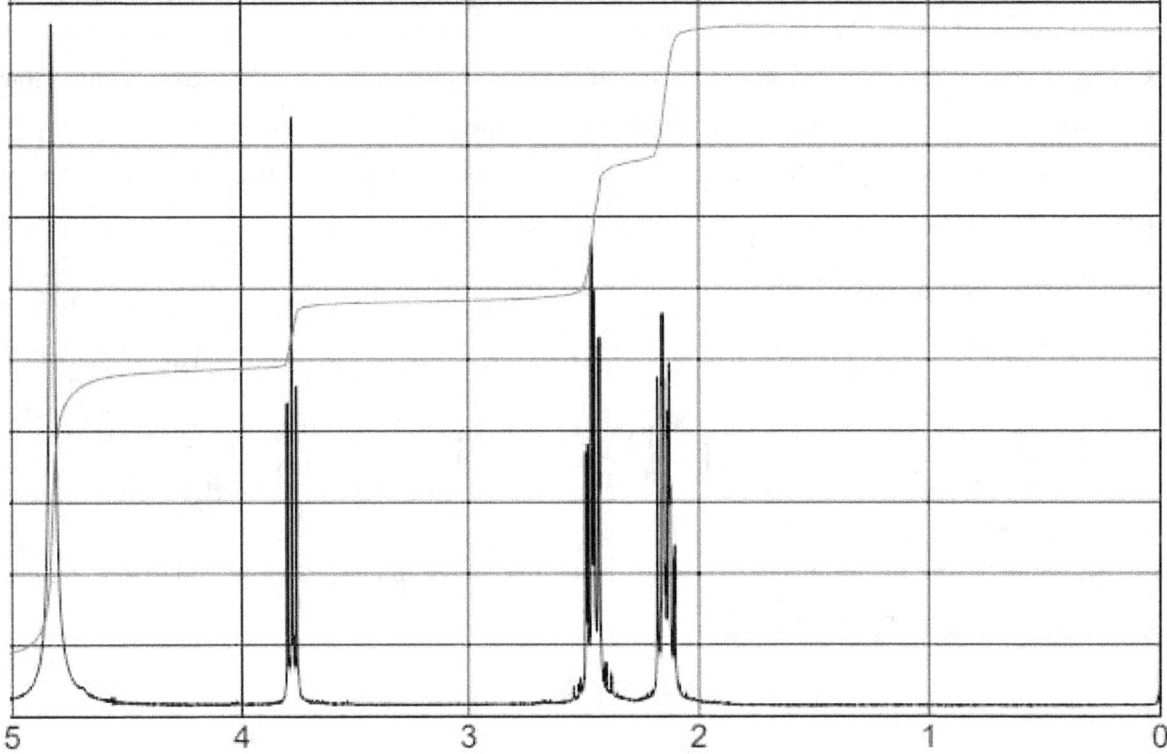

We can immediately discard Candidate II as it would produce two doublets but they are not present and so this leaves us with Candidates I and III.

Candidate III will produce a triplet, of integral two, a doublet of triplets of integral two and a triplet of integral one. These do not appear so that structure can also be discounted.

On the other hand, Candidate I will produce a:-

- Triplet, of integral two;
- Doublet of triplets of integral two;
- Triplet of integral one.

and this is exactly what we observe.

The singlet at δ 4.85 ppm can be assigned to the amino group and the hydroxyl group, as usual, does not appear.

Conclusions

Structure:

H₂N ... NH₂ ... OH

Systematic name: L – glutamine and 2,5-diamino-5-oxopentanoic acid.

Chapter IX

L – Arginine

First isolated in 1886 by Ernst Schulze from yellow lupin seedlings, L – arginine's structure remained controversial until it was synthesised, in 1910, by Søren Sørensen who is most famous for his concept of the pH scale.

Interestingly, L – arginine is classed as a semi-essential or conditionally essential amino acid, since its uses depend on the developmental stage and health of individuals. Unborn babies are unable to synthesise it so it is an essential dietary ingredient for pregnant women some of whom take it as a supplement. Healthy people rarely have to take it as a supplement due to the multiple sources in foods and it is also biosynthesised from L – glutamine (Chapter VIII). As with many amino acids, sources of L – arginine include protein-rich foods such as meat, dairy products and eggs as well as beans, grains and all types of nuts.

L – arginine is the immediate precursor of nitrous oxide (NO) which is an important signalling molecule and is important in the immune system. It is also a biological precursor for urea and is essential for the biosynthesis of creatine (which is used by, amongst others, body builders). It has been found to be medically useful for burns victims, those suffering from sepsis as well as those with kidney failure.

L – arginine has a melting point of 260°C and a boiling point of 368°C and rotates plane polarised light by + 27°.

With a **formula mass** (M_r) of 174.20 g mol^{-1}, this compound has the **elemental composition**: C: 41.33%, H: 8.12 %, N: 32.16 %, O: 18.37 %

This means that L – arginine has the empirical formula of $C_3H_7N_2O$ whilst its molecular formula is $C_6H_{14}N_4O_2$.

L - Arginine

Infrared Spectrum

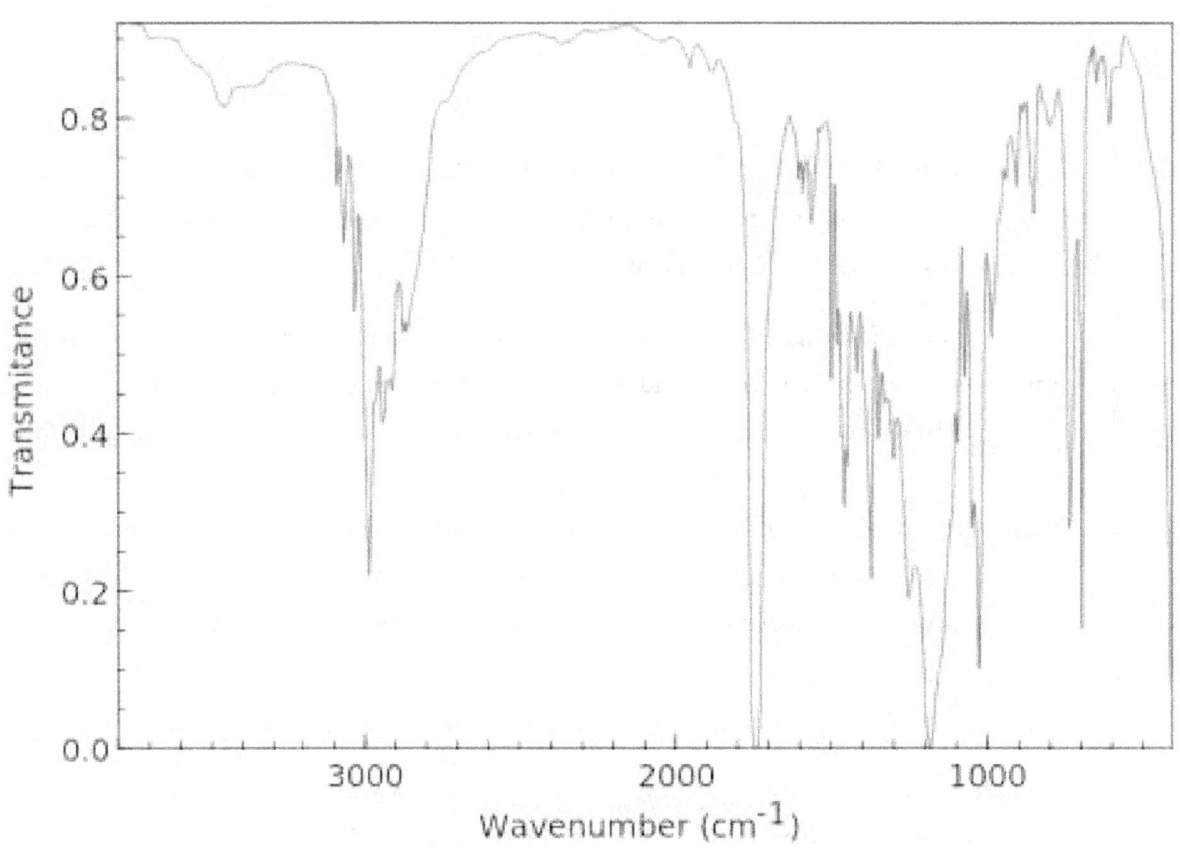

Observations

(√ / X)	Wavenumber range (cm⁻¹)	Wavenumber (cm⁻¹)	Assignment
√	3200 - 3700	3500	O – H
√	3200 - 3600	3500	N – H
X	3000 – 3300		C – H (aromatic)
√	2500 – 3000	3100 - 2900	C – H (aliphatic)
X	2200 – 2500		C ≡ N
?	1700 – 1800	1730	C = O
?	1700 – 1800	1730	C = N
X	1600 – 1700		C = C (aliphatic)
X	1585 – 1600		C – C (aromatic)
X	1450 – 1600		C – C (aromatic)
√	1000 – 1300	1195	C – O
X	700 – 1000		C – X (X = Cl, Br or I)

Conclusions

■ This molecule has insufficient carbon atoms to be aromatic but there does appear to be an N – H or and O – H bond or both. There is also no evidence from this spectrum for the molecule to be aromatic.

■ The compound clearly possesses a carboxylic, –CO$_2$H, functional group as demonstrated by the peaks at 1730 cm^{-1} and 1195 cm^{-1}.

Mass Spectrum

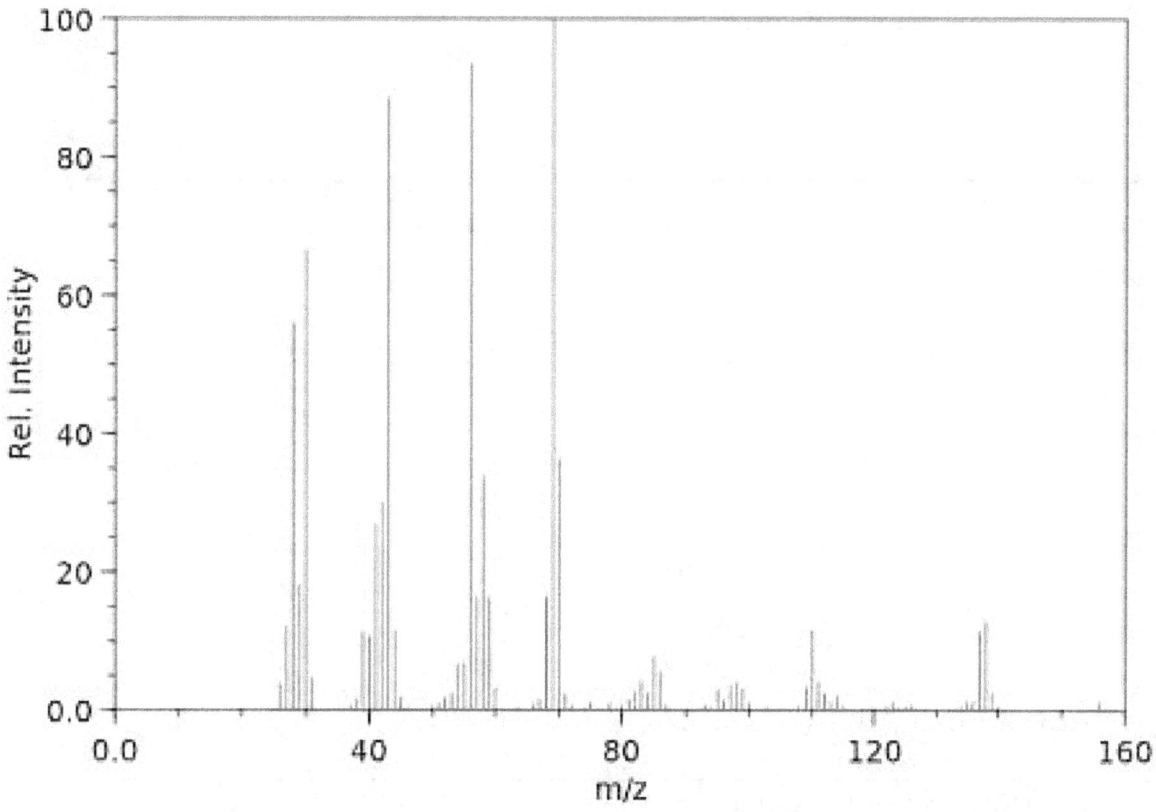

Observations

Charged fragments (m/z)	Assignment	Charged fragments (m/z)	Assignment
Molecular ion: 157	$[C_5H_9N_4O_2]^+$	Base peak: 69	$[C_5H_9]^+$
110	$[C_7H_{10}O]^+$	30	$[CH_2NH_2]^+$
56	$[C_4H_8]^+$	28	$[C_2H_4]^+$

Conclusions

This molecule is either a long chain hydrocarbon, a shorter, substituted compound or an aliphatic ring compound. There are dozens of possible isomers of this compound and they include the following:

▪ Long chain:

There are a number of possible isomers of this molecule alone as the position of the amino, $-NH_2$, group and the $-NH-$ group can be varied.

▪ Substituted:

We can learn much more from the ^1H and ^{13}C NMR spectra.

NMR Spectra

The ^{1}H and ^{13}C NMR spectra are displayed below:

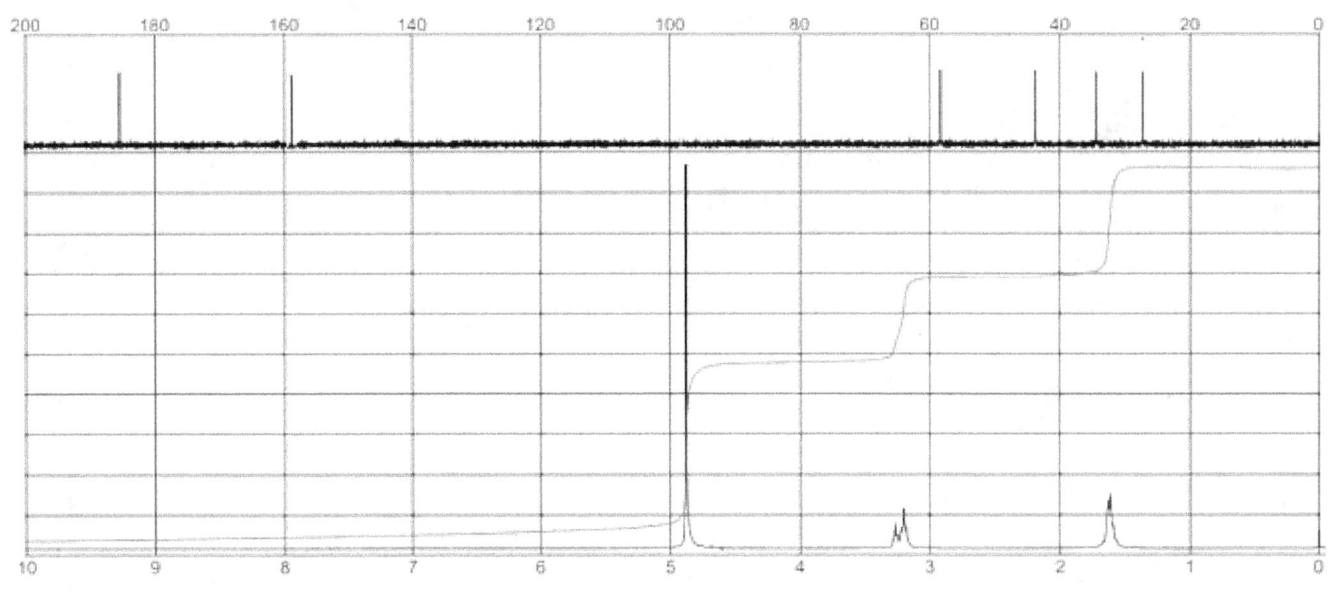

Observations

We will examine the ^{13}C NMR spectrum first.

Chemical shift δ (ppm)	Assignment(s)	Integral
185	C = O or C = N	1
159	C = O or C = N	1
58	C – N	1
43	C – N	1
34	C – C	1
26	C – C	1

We can draw some rapid conclusions but two assignments are tentative:-

- There are two singlets in the C = O / C = N region (δ 185 ppm and δ 159 ppm) so it is reasonable to deduce the presence of a carboxylic acid, –CO₂H, group which must be a terminal group i.e. on one end or the other of a molecule.

- The second peak (δ 159 ppm) is just 1 ppm outside the, guideline, C = O / C = N range. The molecule cannot be aromatic as there is no evidence for aromaticity in the infrared spectrum and so this peak must be due to a C = N group which is likely to exist as a C = NH group.

- The peaks at δ 58 ppm and δ 43 ppm must be due to C – N bonds.

- The presence of the two C – C bonds also leads us reasonably to predict that the molecule contains a four carbon alkane chain to which the carboxylic acid functional group is attached.

As a first proposal, we can can draw the following structure where as usual the squiggles represent so far undetermined groups:

This cannot be correct since it contains too many carbon atoms and so a better guess is:

This does not account for the other two nitrogen atoms or the other two hydrogen atoms so it is reasonable to assume that these constitute two amino, $-NH_2$, groups.

The next problem, however, is that if they are substituents on the carbon chain then they each, individually, replace a hydrogen atom and these two displaced hydrogen atoms must then also be accommodated which is impossible.

The only rational explanation is that one of the these nitrogen atoms exists as an amino, $-NH_2$, substituent but, more importantly, the other nitrogen atom actually exists within the chain. This gives us five possible isomers:

These structures only account for one of the two remaining nitrogen atoms but to try to mark, with squiggles, the possible positions of the amino, $-NH_2$, would make the structures so messy as to be make them unintelligible. It is better to refer back to the ^{13}C NMR spectrum and this allows us to eliminate some candidates immediately.

There are two signals which can be assigned to $C - N$ bonds. One of those will be the bond between the carbon atom and the amino substituent and so there can only be one $C - N$ bond in the chain. This immediately eliminates candidates II, III and IV as they contain, *within the chain*, two $C - N$ bonds leaving none for the $C - NH_2$ substituent where C is a carbon atom in the chain.

We are, therefore, down to candidates I and V whose structures are repeated again below.

We can distinguish between these on the basis of electronegativity. Nitrogen and oxygen are two of the most electronegative elements and are rated at 3.0 and 3.5 respectively on the Pauling electronegativity scale.

* Candidate I provides for two distinct signals in the $C = O$ / $C = N$ region which is what we observe.
* In contrast, candidate V has two groups containing a nitrogen atom and so the signals will appear extremely close together. This is not observed and so we can select Candidate I as the structure.

We now need to determine the position of the substituted –NH₂ group and we have four possibilities as shown below. To prevent any confusion these structures are labelled alphabetically.

Referring back, again, to electronegativity, Candidates B and C make no sense as the amino group is too far away from the terminal groups to have any deshielding effect and that leaves us with candidates A or D as the correct structure. Given the similar electronegativities of nitrogen and oxygen, a difference of 3.0 and 3.5 is not enormous, Candidate A would produce two signals in the δ 160 – 220 ppm region which are very close together.

In contrast, the amino group will deshield the carbon atom in the carboxylic acid group producing two signals significantly apart from each other and so this must be the correct structure.

¹H NMR Spectrum

We can label the hydrogen atoms as shown below even though the ¹H NMR spectrum provides relatively little useful information.

but the following can be noted:

- The broad, ill-defined multiplet at δ 1.62 ppm of integral four must be due to C – H hydrogen atoms which are highlighted by the rectangle;
- The two multiplets at δ 3.25 and 3.10 ppm of integrals two and one respectively can be assigned to Ha/Ha' and Hb respectively.

Conclusions

Structure:

Systematic name: L – arginine and 2-amino-5-guanidinopentanoic acid.

45

Chapter X

L – Lysine

First isolated, in 1889, from milk by Ferdinand Drechsel the structure of L – lysine was determined in 1902 by the famous chemist, Emil Fischer

L – lysine is important in a number of processes including the production of protein, the production of carnitine which is used in the metabolism of fatty acids but also, perhaps most interestingly, the uptake of essential mineral nutrients.

It is an essential amino acid and L – lysine is found in all protein-containing foods such as dairy products, meats, soy and beans as well as some fish such as cod.

L – lysine has a melting point of 215°C and a boiling point of 312°C and it rotates the angle of plane polarised light by +15°.

With a **formula mass** (M_r) of 146.19 g mol^{-1} , this compound has the **elemental composition**: C: 49.25 %, H: 9.67%, N: 19.16%, O: 21.89% which gives it an empirical formula of C_3H_7NO and a molecular formula of $C_6H_{14}N_2O_2$

L – Lysine

Infrared Spectrum

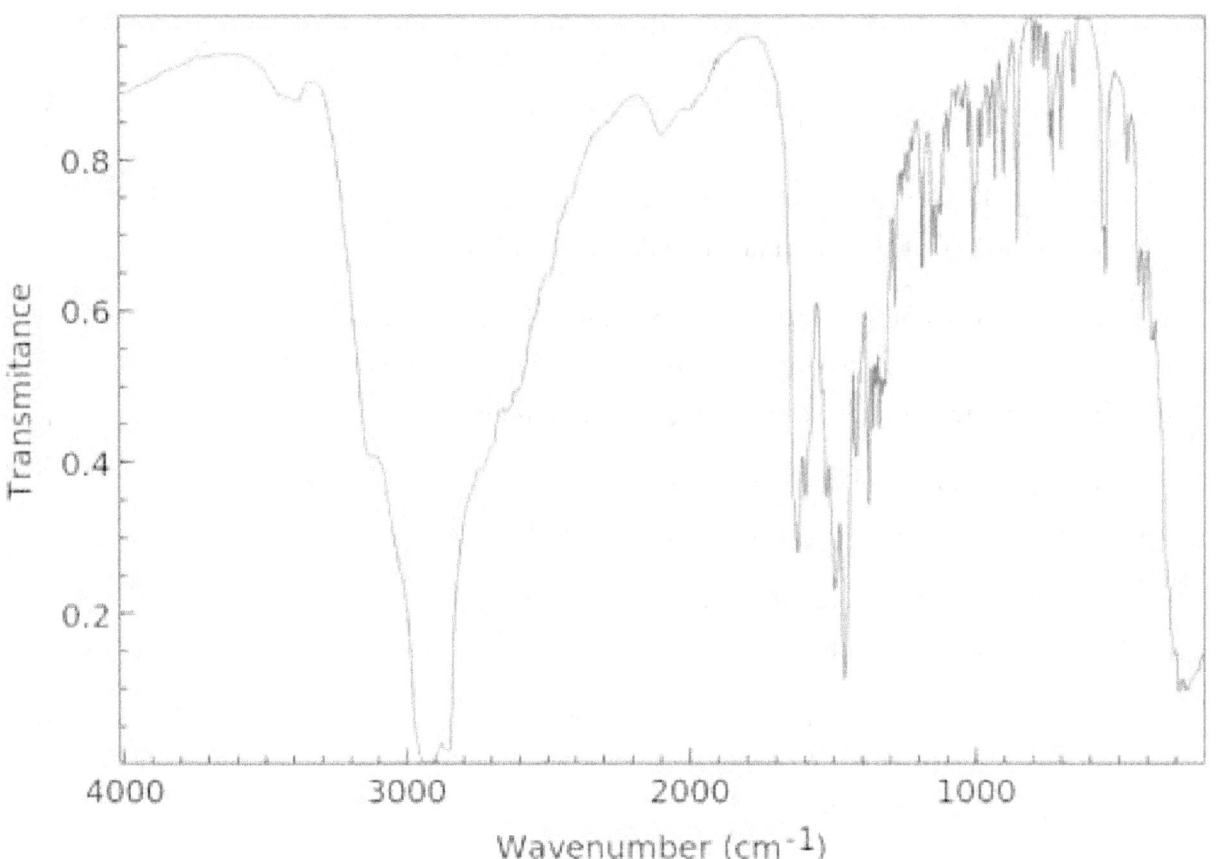

Observations

(√ / X)	Wavenumber range (cm⁻¹)	Wavenumber (cm⁻¹)	Assignment
?	3200 - 3700	3350, 3150	O – H
?	3200 - 3600	3350, 3150	N – H
X	3000 – 3300		C – H (aromatic)
X	2500 – 3000	2950, 2850	C – H (aliphatic)
X	2200 – 2500		C ≡ N
X	1700 – 1800		C = O
X	1700 – 1800		C = N
√	1600 – 1700	1650	C = C (aliphatic)
X	1585 – 1600		C – C (aromatic)
X	1450 – 1600		C – C (aromatic)
√	1000 – 1300	Fingerprint region	C – O
X	700 – 1000		C – X (X = Cl, Br or I)

Conclusions

It is difficult to draw many conclusions from this spectrum beyond stating that the molecule appears to be aliphatic, contains O – H and N – H groups and may contain a C = C bond. It is possible, though, that, since the molecule contains two oxygen atoms, that the peak at 1650 cm⁻¹ actually demonstrates the presence of a carboxylic acid, – CO_2H, group.

We can learn more from the mass and ¹H and ¹³C NMR spectra.

Mass Spectrum

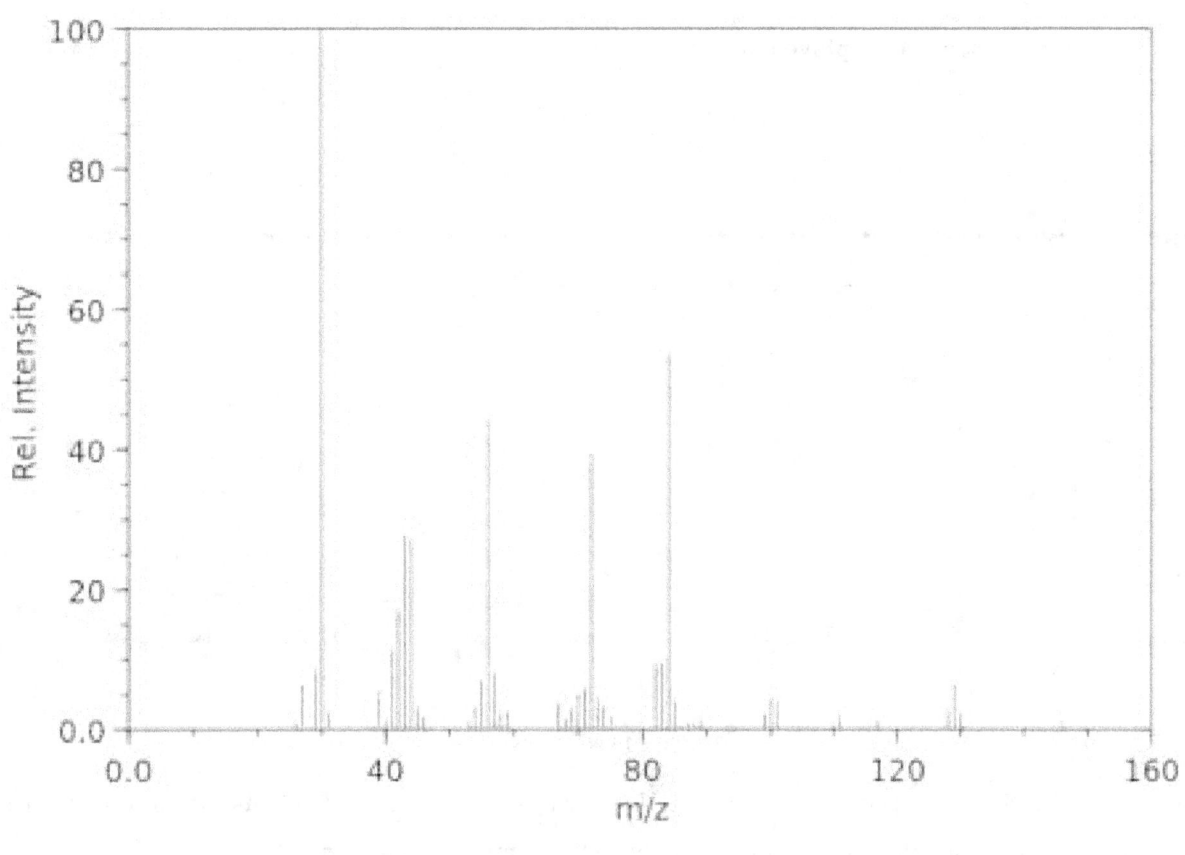

Observations

Charged fragments (m/z)	Assignment	Charged fragments (m/z)	Assignment
Molecular ion: 146	$[C_6H_{14}N_2O_2]^+$	Base peak: 30	$[CH_2NH_2]^+$

84	$[C_6H_{12}]^+$	45	$[CO_2H]^+$
72	$[C_2H_5\text{-}CO\text{-}CH_2+H]^+$	43	$[C_3H_7]^+$
56	$[C_4H_8]^+$	28	$[C_2H_4]^+$ / $[CNH_2]+$

Conclusions

❦ This molecule contains a six carbon chain with an amino, $-NH_2$, group.

❦ The peak at m/z = 45, provides evidence for the existence of a carboxylic acid, $-CO_2H$, group.

If this is correct then we have the basic chain as shown below:

The remaining amino, $-NH_2$, group indicated by the $+NH_2$ in the diagram above must be a substituent somewhere on the carbon chain and so we have five possible candidate isomers for this molecule which are displayed in the next section when we consider the ${}^1H$ and ${}^{13}C$ nmr spectrum which is our final task.

NMR Spectra

The ¹H and ¹³C nmr spectra are displayed below:

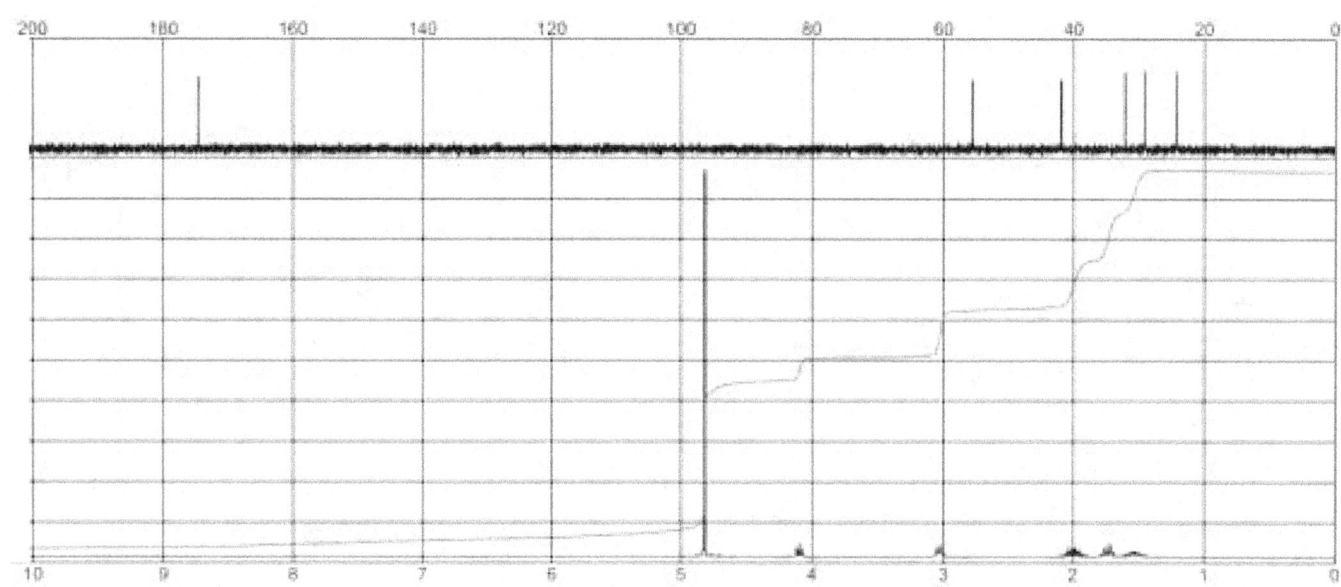

Observations

We will consider the ¹³C nmr spectrum before the ¹H nmr spectrum but our first task is to consider the five possible isomers of the molecule discussed previously. The isomers are shown below:

¹³C NMR Spectrum

In the ¹³C nmr spectrum we observe the following:

Chemical shift δ (ppm)	Assignment(s)	Integral
175	C = O	1
56	C – N	1
43	C – N	1
32	C – C	1
29	C – C	1
25	C – C	1

and we can draw the following conclusions:-

▪ The peak at δ 175 ppm (C=O) demonstrates that the molecule contains a carboxylic acid, –CO₂H, group which must exist on one end of the molecule;

▪ The peaks at δ 56 ppm and δ 43 ppm can be assigned to C – O and C – N bonds but as the regions listed in the data sheet overlap it is not possible to be completely certain which of the two peaks is which but one must be due to C – O and the other due to C – N;

▪ There is no evidence for the presence of a C = C bond as there is no signal in the δ 110 – 160 ppm region and so the 1650 cm⁻¹ peak in the infrared region must be due to a C = O bond and not as originally suggested;

▪ The peak at δ 43 ppm must be due to a C – N bond;

▪ The peaks at δ 32 ppm, δ 29 ppm and δ 25 ppm can all be assigned to alkyl carbon atoms.

This means that we can discard candidate molecules I, III and IV since candidate I contains three C – C bonds whilst candidates III and IV contain only one C – C bond leaving us with:

as possible candidates. The C – C bonds are highlighted with ovals and we can distinguish between the two isomers using the ¹H nmr spectrum. Before we detail the peaks and multiplicities it is useful to predict the spectra and we will do so in the next section where we label the hydrogen atoms alphabetically.

¹H NMR Spectrum

The labelled molecules are shown below and we will, in turn, predict the chemical shifts, integrals and multiplicities of each hydrogen atom but we will not consider the hydroxyl, –OH, or the amino, –NH₂, hydrogen atoms as, despite being labelled, they often do not appear in the ¹H nmr spectrum whilst, when they do they can appear anywhere in the δ 0 – 12 ppm region. We will consider each in turn.

Candidate II:

▪ H_b will produce a doublet of integral two due to the splitting by H_c;

▪ H_c will generate a doublet of triplets, of integral one, due to the signal being split by H_b and H_e;

▪ H_e will cause a doublet of triplets , of integral two, due to splitting by H_f and then H_c;

▪ The signal due to H_f will be a quintet due to splitting by the two H_e and the two H_g hydrogen atoms;

▪ H_g's signal will be a triplet of integral two, due to splitting by H_f.

If signals due to the four (two pairs) –NH₂ hydrogen atoms do appear then, since there is little difference between the two groups, we will observe a singlet of integral four. It will be clear what causes it.

Candidate V:

Again without considering H_i, H_o and H_p and remembering that H_j, H_k, H_l and H_m represent two hydrogen atoms whilst H_n represents one hydrogen atom:

- H_j will produce a triplet of integral two due to splitting by H_k;

- H_k will cause a doublet of triplets, of integral two, due to splitting by H_j and H_l;

- The signal due to H_l will be a quintet, of integral two, due to splitting by H_k and H_m;

- H_m will generate a doublet of triplets, of integral two, due to splitting by H_l and then H_n;

- H_n will generate a triplet, of integral one, due to splitting by H_m.

We need to examine the enlarged ^1H nmr spectrum which is shown below.

In the spectrum, two of the signals are complex multiplets and impossible to analyse but we only need to examine the triplets at δ 3.05 ppm (integral two) and at δ 4.18 ppm (integral one) and the quintet of integral two at δ 1.75 ppm.

These triplets could only have been generated by candidate V as it is the only one that we predicted would produce three distinct multiplets.

Conclusions

Structure:

Systematic name: L – lysine and (2S)-2,6-diaminohexanoic acid.

Chapter XI

L – Methionine

First isolated in 1921 by John Howard Mueller, L – methionine is an essential amino acid in humans. Since it cannot be synthesised by the human body it must be consumed as part of a balanced diet and it is found in many foodstuffs including fish, meat and dairy products as well as brazil nuts.

L – methionine is involved in the formation of new blood vessels but is also metabolised by the body to produce other amino acids such as L – cysteine (the only other sulfur-containing amino acid- Chapter II). It is also used by the body to produce taurine, a major component of bile, and glutathione which is an important antioxidant.

Used as a treatment for liver disorders, L – methionine is believed to prevent the metabolised products of acetaminophen from causing further liver damage. It is also an effective anti – viral medication and is used as an antidote for copper poisoning however it has been suggested that it may also act in the growth of cancerous cells. This is interesting since it also has anti-oxidant properties so may also kill cancer cells.

L – methionine has a melting point of ~281°C at which temperature it decomposes and in 1M hydrochloric acid, rotates plane polarised light by +23°.

With a **formula mass** (M_r) of 149.21 g mol^{-1} , L – methionine has the **elemental composition** C: 40.21 %, H: 7.45 %, N: 9.39%, O: 21.45 %, S: 21.45 % meaning that the empirical and molecular formulas are both $C_5H_{11}NO_2S$.

Infrared Spectrum

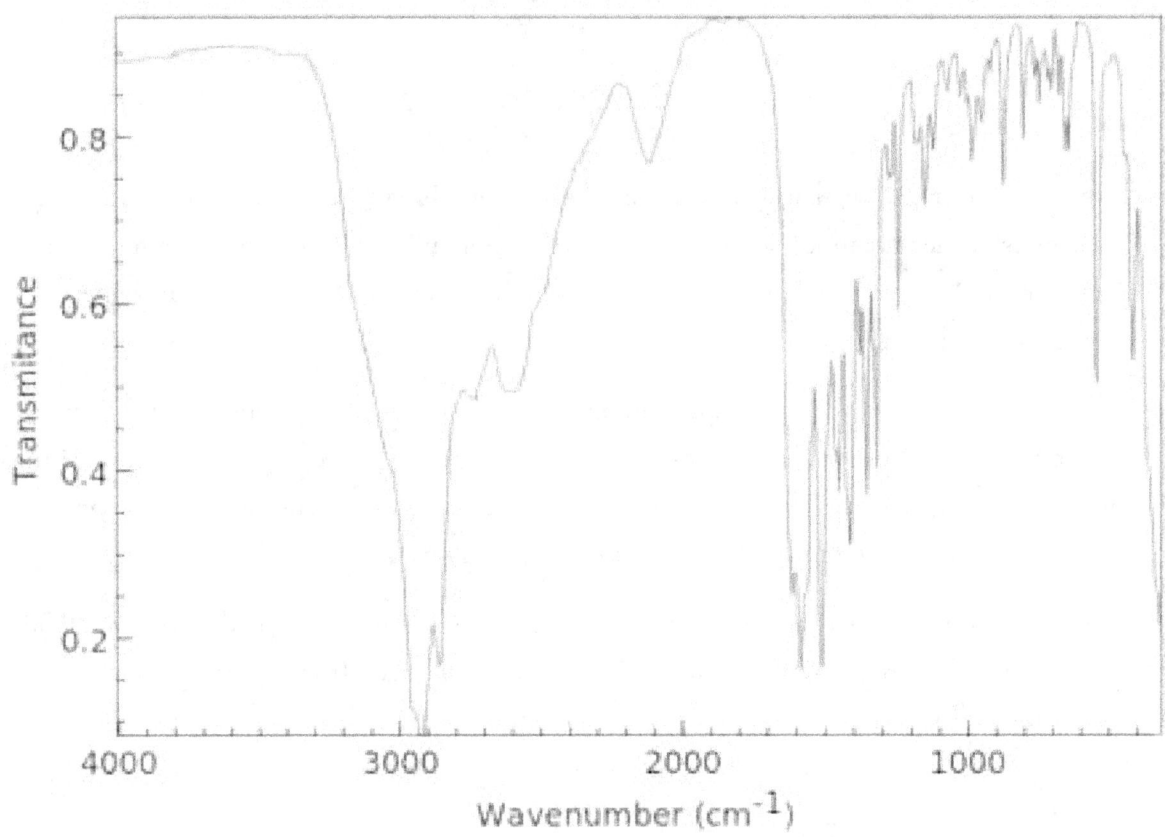

Observations

(√ / X)	Wavenumber range (cm⁻¹)	Wavenumber (cm⁻¹)	Assignment
?	3200 - 3700	3210	O – H
?	3200 - 3600	3210	N – H
X	3000 – 3300		C – H (aromatic)
√	2500 – 3000	2920, 2820	C – H (aliphatic)
X	2200 – 2500		C ≡ N
X	1700 – 1800		C = O
X	1700 – 1800		C = N
√	1600 – 1700	1620, 1600	C = C (aliphatic)
X	1585 – 1600		C – C (aromatic)
X	1450 – 1600		C – C (aromatic)
√	1000 – 1300	Fingerprint region	C – O
X	700 – 1000		C – X (X = Cl, Br or I)

Conclusions

This spectrum is confusing and of very limited use but we can make some general observations:

- There is a broad stretch to the left of 3000 cm⁻¹ which may indicate the presence of an O – H or an N – H bond or both;

- There appears to be a C = C bond;

- There appears to be no carboxylic acid, –CO₂H, group.

Mass Spectrum

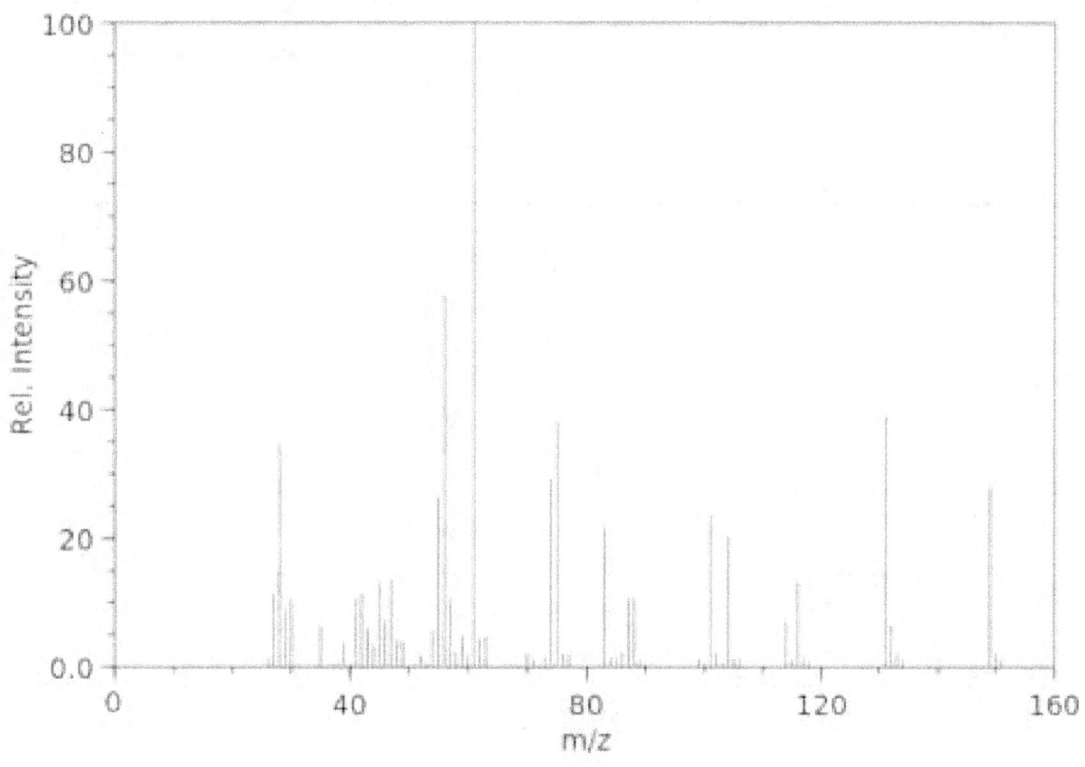

Observations

Charged fragments (m/z)	Assignment	Charged fragments (m/z)	Assignment
Molecular ion: 149	$[C_5H_{11}NO_2S]^+$	Base peak: 61	$[CH_3COO+2H]^+$

Charged fragments (m/z)	Assignment	Charged fragments (m/z)	Assignment
131	$[C_4H_9NO_2S]^+$	75	$[C_3H_7O_2]^+$
116	$[C_4H_8O_2S]^+$	56	$[C_4H_8]^+$
101	$[C_3H_5O_2S]^+$	28	$[C_2H_4]^+$

Conclusions

This spectrum contains a lot of fascinating information but we do not need to consider all of it.

It is clear that:-

※ Contrary to the infrared spectrum, the molecule contains a carboxylic acid, $-CO_2H$, group and that;

※ It also contains a four carbon chain whilst;

※ The peaks at m/z = 101 and 116 imply that the sole sulfur atom is also in the chain;

※ The molecular ion (m/z = 149) contains a nitrogen atom indicating that either:

  ※ The other end of the molecule to the carboxylic acid group terminates in an amino, NH_2, group or;

  ※ If, the sulfur atom is in the chain, the chain terminates in a methyl, $-CH_3$, group.

This gives us a number of possible candidates where the sulfur atom may be in different positions in the chain.

We can determine the structure from consideration of the ^1H and ^{13}C nmr spectra which is our next task.

NMR Spectra

The ¹H and ¹³C nmr spectra are displayed below:

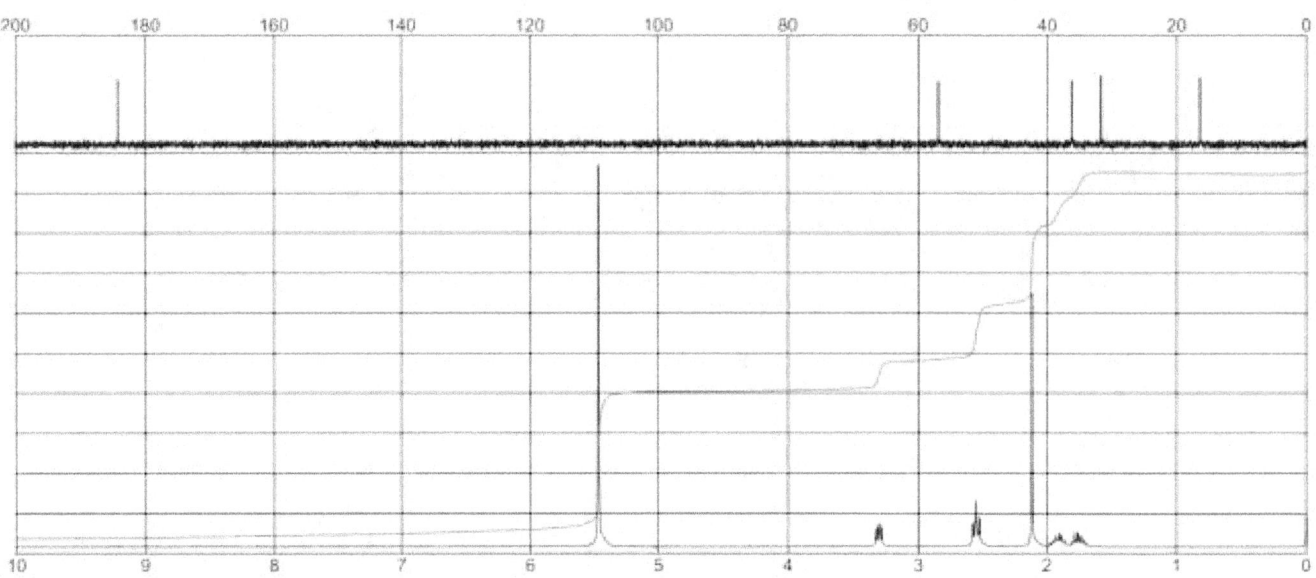

We will consider the ¹³C nmr spectrum first.

Observations

Chemical shift δ (ppm)	Assignment(s)	Integral
184	C = O	1
57	C − N	1
35	C − N	1
32	C − S	1
17	C − C	1

Conclusions

▪ From the signal at δ 184 ppm, the molecule contains a carboxylic acid, –CO₂H, functional group;

▪ The peak at δ 57 ppm can be assigned to a C − N bond whilst;

▪ The singlets at δ 35 and δ 32 ppm must be due to C − S bonds and, since, there is only one sulfur atom in the molecule, it must be in the chain;

▪ The peak at δ 17 ppm is due to an alkyl carbon atom..

We can begin to construct the molecule starting with the carboxylic acid group with the C − C bonded to it and the basic chain. The sulfur atom forms part of the chain. Without considering the position of the amino group, there are four possible base isomers:

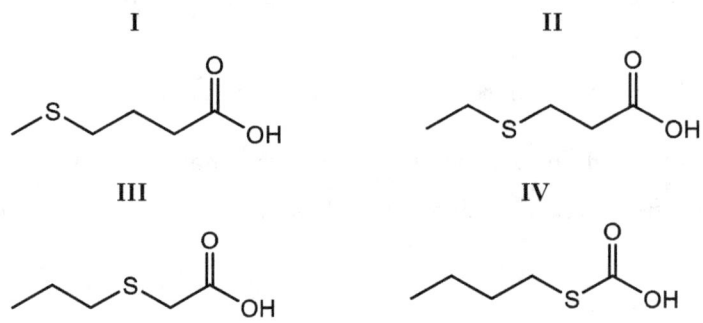

but we can eliminate candidates II, III and IV since candidates II and III both contain two C – C bonds and candidate IV contains three C – C bonds whereas the ^{13}C nmr spectrum indicates that the molecule contains only one. This leaves us with candidate I only and we now need to determine the position of the amino, –NH$_2$, group. This gives us three possibilities which are shown below labelled sequentially:

We can immediately dispense with:-

* Candidate V as it contains three carbon atoms forming aliphatic C – C bonds. This is not observed in the ^{13}C nmr spectrum;

* Candidate VI as the molecule does not contain any aliphatic C – C bonds; each of the non-carboxylic acid carbon atoms forms C – N or C – S bonds and this leaves us with Candidate VII only.

We can prove or disprove the hypothesis by predicting and then examining the ^{1}H nmr spectrum.

^{1}H NMR Spectrum

We will consider the proposed candidate molecule we can predict the ^{1}H nmr spectrum before examining the actual spectrum. The candidate structure with labelled hydrogen atoms is shown below:

It is important to note that the two hydrogen atoms on the carbon atom are not chemically and magnetically equivalent due to their proximity to the chiral H$_d$ hydrogen atom and are labelled as H$_c$ and H$_{c'}$. This is due to the amino nitrogen atom being tetrahedral due to the existence of a lone pair of electron on the nitrogen atom. We can make the following predictions:

* H$_a$ will produce a singlet, of integral three, in the region δ 4 – 6 ppm;

* The signal due to the two H$_b$ hydrogen atoms will be a triplet, of integral, two due to splitting by the two hydrogen atoms, H$_c$ and H$_{c'}$.

* The signals due to H$_c$ and H$_{c'}$ are slightly more complicated.

  H$_c$ will be split into a triplet by H$_b$ which is split into a sextet by H$_{c'}$ The resultant signal will have an integral of one.

  H$_{c'}$ will be split in a similar manner.

* H$_d$ will result in a triplet, of integral one, due to splitting by H$_c$ which is then split into a doublet of triplets.

L - Methionine

All these signals can be observed in the expanded ¹H nmr spectrum.

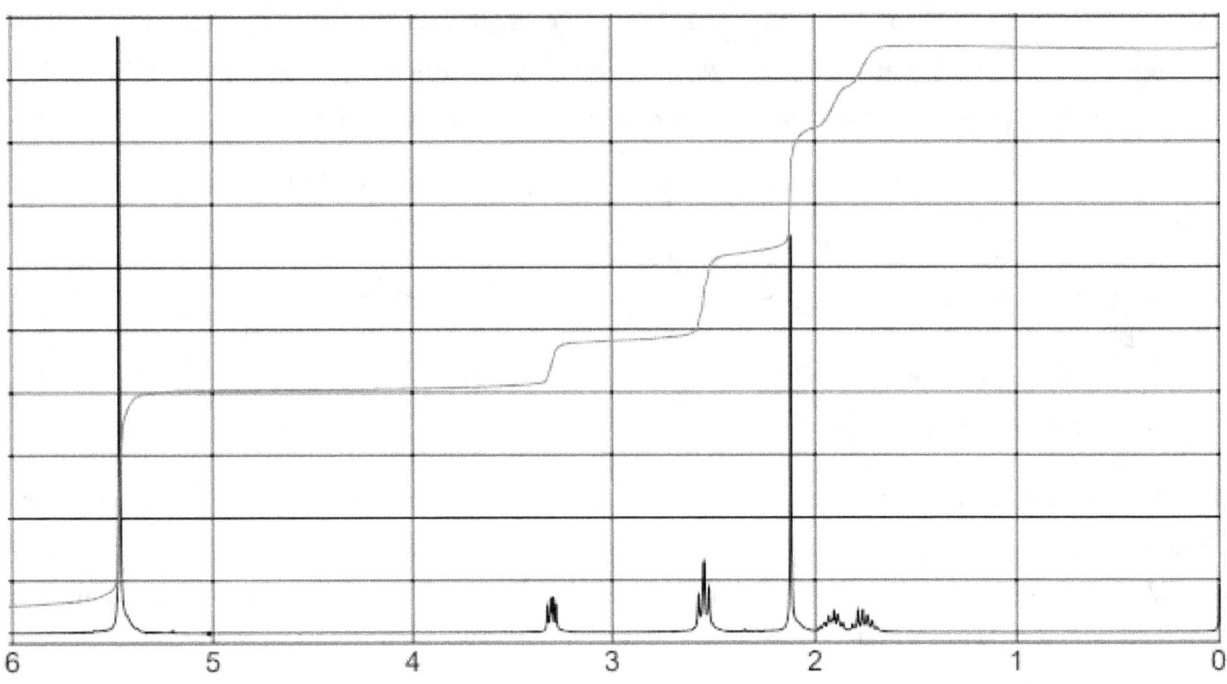

and the assignments are as follow:

Chemical shift δ (ppm)	Integral	Multiplicity	Assignment
5.45	3	Singlet	H_a
3.35	1	Triplet	H_d
2.60	2	Triplet	H_b
2.15	2	Singlet	H_e
1.90	1	Sextet	H_c or $H_{c'}$
1.70	1	Sextet	H_c or $H_{c'}$

Conclusions

Structure:

Systematic name: L-methionine and 2-amino-4-(methylthio)butanoic acid.

Chapter XII

L – Phenylalanine

First isolated in 1879 by Ernst Schulze and Johann Barbieri from the seedlings of yellow lupines, L – phenylalanine was synthesised three years later by Richard August Carl Emil Erlenmeyer (who is most famous for his design of the Erlenmeyer flask which is familiar to all chemists) from phenylacetaldehyde, hydrogen cyanide and ammonia.

L – phenylalanine is an essential amino acid and is used by the body to both form proteins but is also the precursor to L – tyrosine, dopamine, noradrenaline (norepinephrine) and melanin. It occurs naturally in breast milk and is also used in the industrial production of many foods and drinks. It is also marketed as a nutritional and dietary supplement and is claimed to have both analgesic and antidepressant properties. As an essential amino acid, it must be incorporated into a balanced diet and it is found in dairy products, beef and soy beans and can also be acquired from any food product containing the artificial sweetener, L – aspartame which metabolises, through hydrolysis, to form L – phenylalanine, L – aspartic acid and methanol.

L – phenylalanine has a melting point of about 270°C and a boiling point of about 320°C and rotates plane polarised light by -33°. With a **formula mass** (M_r) of 165.19 g mol^{-1}, it has the **elemental composition**: C: 65.38 %, H: 6.72 %, N: 8.48 %, O: 19.37 % which means that it has the empirical and molecular formulas $C_9H_{11}NO_2$.

Infrared Spectrum

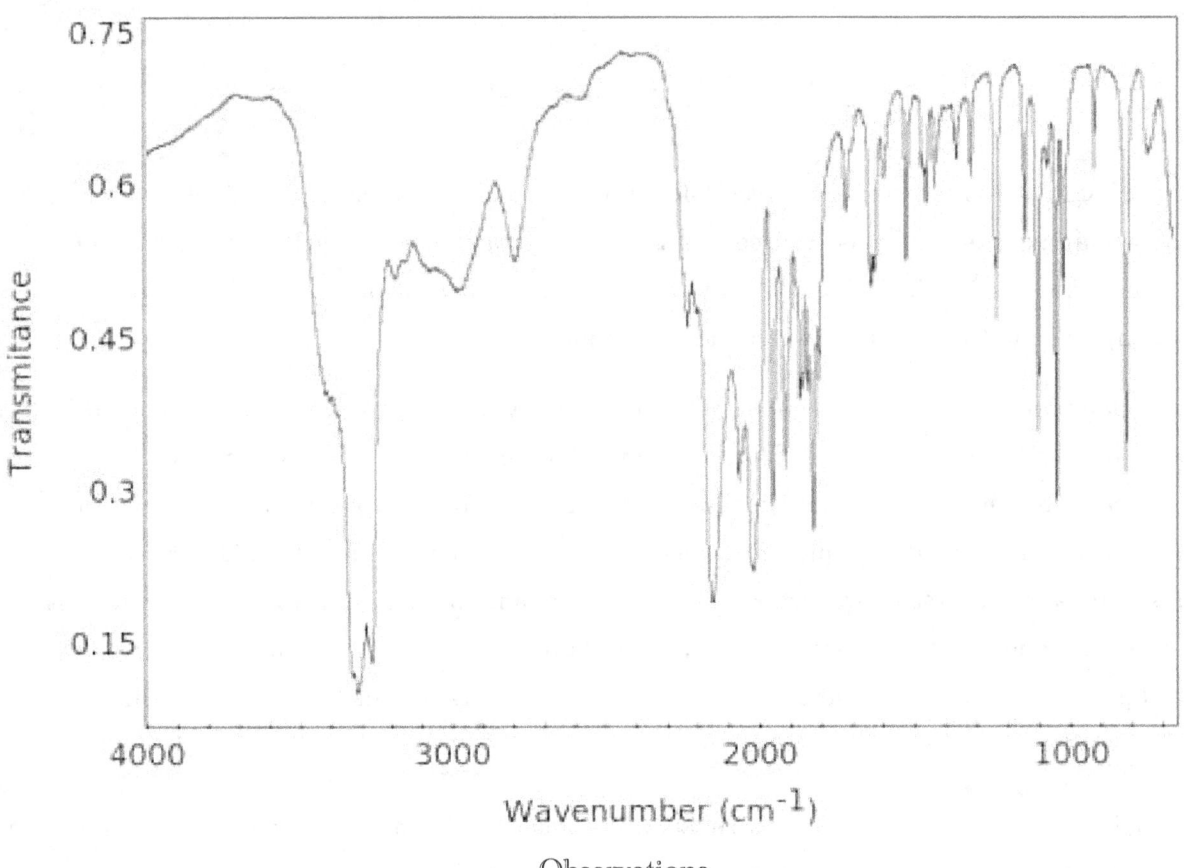

Observations

(√ / X)	Wavenumber range (cm⁻¹)	Wavenumber (cm⁻¹)	Assignment
?	3200 - 3700	3450	O – H
?	3200 - 3600	3450	N – H
√	3000 – 3300	3290	C – H (aromatic)
√	2500 – 3000	2900, 2700	C – H (aliphatic)
X	2200 – 2500		C ≡ N
?	1700 – 1800	1700	C = O
?	1700 – 1800	1700	C = N
X	1600 – 1700		C = C (aliphatic)
X	1585 – 1600		C – C (aromatic)
X	1450 – 1600		C – C (aromatic)
√	1000 – 1300	Fingerprint region	C – O
X	700 – 1000		C – X (X = Cl, Br or I)

Conclusions

This spectrum is difficult to interpret as the stretch to the left of 3000 cm⁻¹ appears to demonstrate the presence of aromatic hydrogen. There appears to be an overlap of peaks which may be caused by O – H and/or N – H groups whilst the peak at 1700 cm⁻¹ also indicates the presence of a carboxylic acid group.

Mass Spectrum

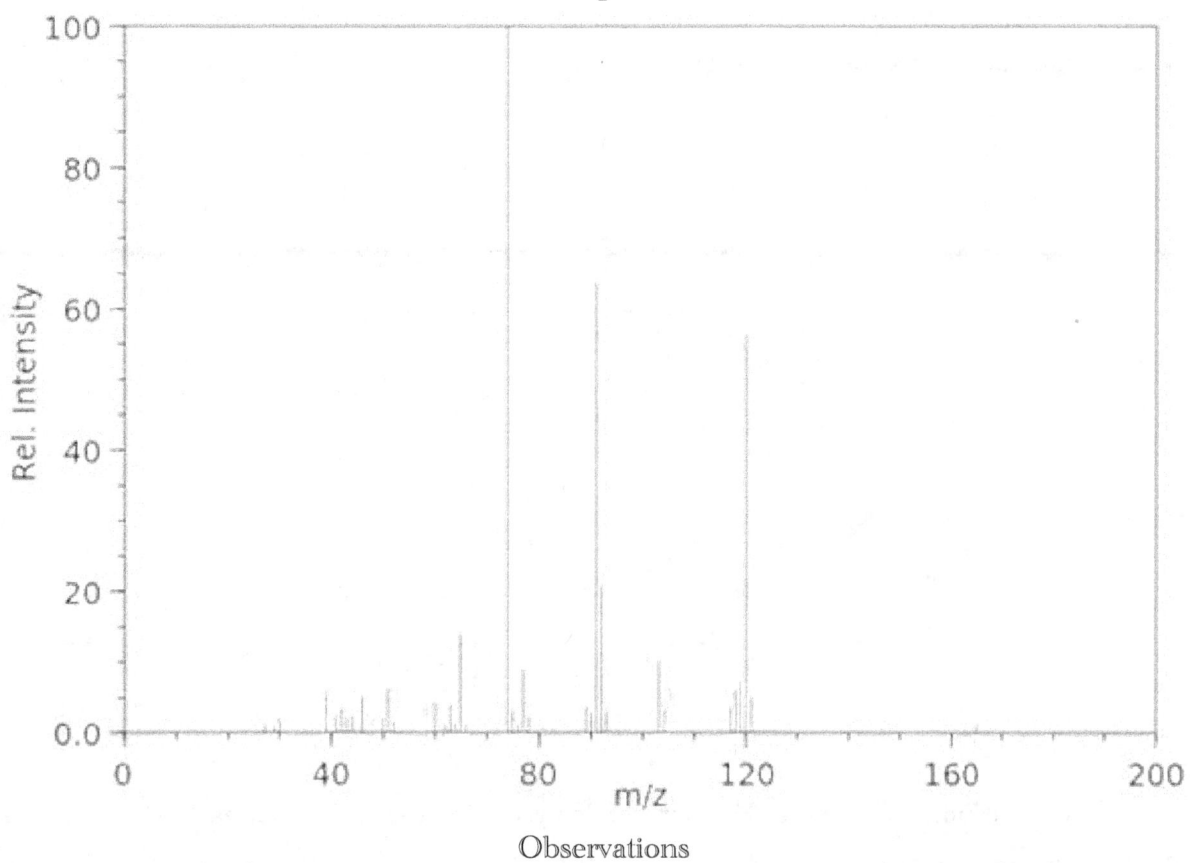

Observations

Charged fragments (m/z)	Assignment	Charged fragments (m/z)	Assignment
Molecular ion: 165	$[C_9H_{11}NO_2]^+$	Base peak: 74	$[C_3H_6O_2]^+$ / $[C_2H_4NO_2]^+$
120	$[C_6H_5CH_2CHNH_2]^+$	77	$[C_6H_5]^+$
103	$[C_6H_5CH_2C]^+$	39	$[C_3H_3]^+$
91	$[C_6H_5CH_2]^+$	30	$[C_2NH_2]^+$

Conclusions

The peak at m/z = 77 is characteristic of a phenyl ring which is a benzene ring with one of the hydrogen atoms replaced by a substituent group. It has been observed that a peak at this mass to charge ratio is nearly always due to such a group. This means that we can draw the molecule as:

We know from the infrared spectrum that the molecule also contains a carboxylic acid group which must be on the terminal end of the substituent and this means that there are only two possible structures as shown:

These isomers can be distinguished by examination of the ^1H and ^{13}C nmr spectrum.

NMR Spectra

The ^1H and ^{13}C nmr spectra are displayed below:

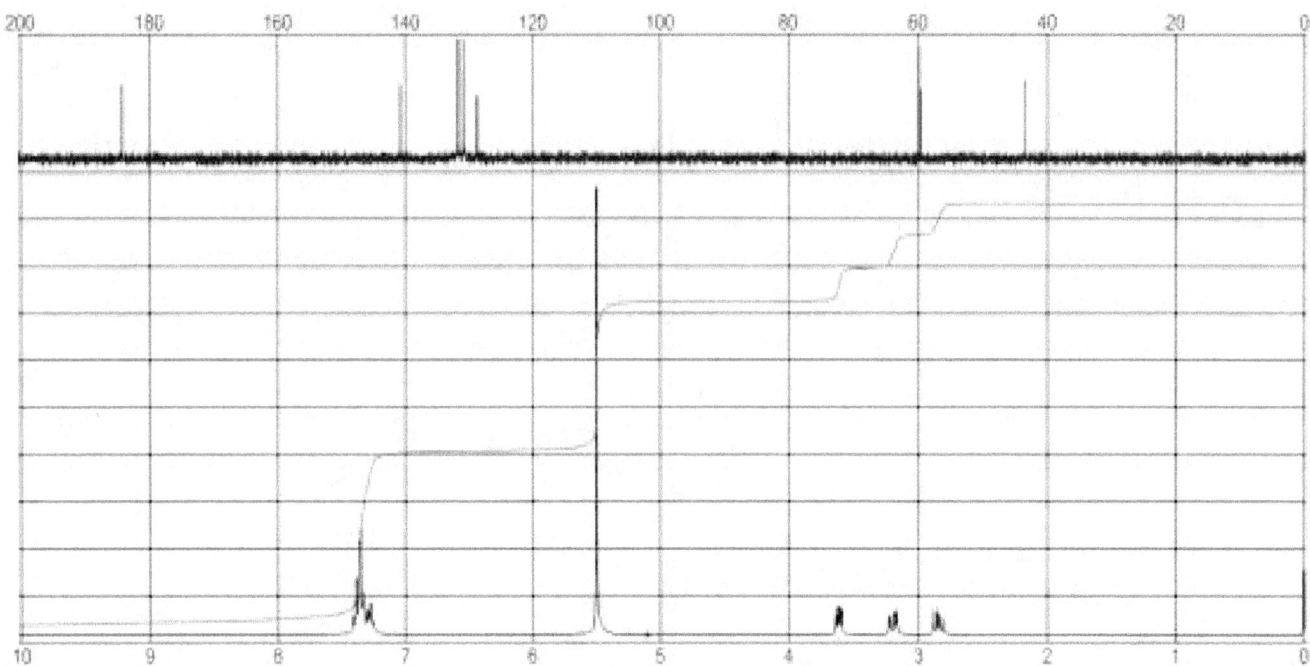

As there are only two candidate isomers it is worthwhile predicting the spectra for each which can can then compare with measured spectrum.

Although we will predict the spectra separately for the two candidate molecules it is noteworthy that the signals due to the phenyl hydrogen atoms which will appear in the δ 7 – 8 ppm region will be the same for the two molecules. If we rotate the molecules it becomes obvious that there should be three signals in the ratio 2:2:1 due to the symmetry of the molecules. This is demonstrated below using candidate I as an example:

There are two chemically and magnetically equivalent ortho- hydrogen atoms, two chemically and magnetically equivalent meta- hydrogen atoms and a single para- hydrogen atom (which is not highlighted in the structure above. Thankfully we only need to consider three of them and we will look at H_a, H_b and H_c but the same predictions would apply to $H_{a'}$, $H_{b'}$ and H_c.

- H_a would be split into a doublet by H_b and this doublet would be split into a doublet of doublets by H_c. This would have an integral of two;
- H_b would be split into a triplet by H_a and H_c. This would also have an inegral of two;
- H_c would be split into a doublet by H_b and then the doublet would be split into a doublet of doublets by H_a. This would have an integral of one.

If we examine the ^{1}H NMR spectrum we can see that there are two multiplets of integral two and one of integral one. It is now time to predict and the analyse the ^{1}H NMR spectra of the side chain and we will do this for each candidate in turn.

^{1}H NMR Spectrum Predictions

Candidate I

Given the planar nature of the molecule, the two hydrogens on the $- CH_2 -$ group attached to the phenyl ring will be chemically and magnetically non-equivalent and we can label the hydrogen atoms as shown below:

We should observe the following:-

▮ H_f will produce a doublet of doublets, of integral one, due to splitting by H_f and then by H_g.

▮ $H_{f'}$ will produce a doublet of doublets, of integral one, due to splitting by H_f and then by H_g.

▮ The signal due to H_g will be a doublet of doublets, of integral one, due to splitting by H_f and then by $H_{f'}$;

▮ There will be a singlet, of integral two, due to H_h;

▮ Due to its lability, there may or may not be a signal assignable to H_i.

Candidate II

We should observe the following:-

▮ The signal due to H_o will be a triplet, of integral one, due to splitting by the two H_q hydrogen atoms;

▮ There will be a singlet, of integral two, due to H_p;

▮ The signal assignable to H_q is similarly complicated to the equivalent atoms in candidate I. This is because the carboxylic acid group is planar and the two H_q hydrogen atoms are also adjacent to the chiral carbon atom which means that these are not equivalent. Relabelling the hydrogen atoms gives us the following structure:

▮ H_q will produce a doublet of doublets, of integral one, due to splitting by $H_{q'}$ and then by H_o.

▮ Similarly, $H_{q'}$ will produce a doublet of doublets, of integral one, due to splitting by H_q and then by H_o

▮ Again, due to its lability, there may or may not be a signal assignable to H_r.

Initially, these spectra appear extremely similar but there is one difference:

Candidate I will produce three doublets of doublets whilst Candidate II will produce two doublets of doublets and a triplet. If we examine the expanded ^{1}H nmr spectrum we observe the presence of three doublets of doublets so the molecule must have this structure:

The ^{1}H nmr spectrum is interesting for another reason. One of the doublet of doublets spans another doublet of doublets centred on δ 3.5 ppm and at low resolution would probably appear as a quartet.

We must finally consider the ^{13}C nmr spectrum.

^{13}C NMR Spectrum

For this purpose it is useful to view the molecule in another orientation as it helps us to see the there are two pairs of chemically and magnetically equivalent aromatic hydrogen atoms. The re-orientated molecule with labelled carbon atoms is shown below:

Observed from this perspective it is clear that C_e and $C_{e'}$ are chemically and magnetically equivalent as are C_f and C_f. This means that we can make the following assignments:

Chemical shift δ (ppm)	Assignment(s)	Integral
184	C_a	1
141	C_d	1
131	C_e / $C_{e'}$	2
130	C_f / C_f	2
129	C_g	1
59	C_b	1
43	C_c	1

Conclusions

Structure:

Systematic name: L – phenylalanine and (S)-2-amino-3-phenylpropanoic acid.

Chapter XIII

L – Proline

First isolated in 1900 by Richard Willstätter, L – proline is used by the body to make proteins such as collagen and is also effective in healing skin damage. It also has a number of roles in the general functioning of cells. It is a non-essential amino acid as the body can produce it from L – glutamine (Chapter VIII) but is also consumed in protein-rich diets such as meat, fish and dairy products.

In food manufacturing, L – proline is used to produce haze or cloudiness in beers whilst in the pharmaceutical industry it is an ingredient in many medicines to prolong their shelf lives.

It has also been found that plants under stress may accumulate L – proline and it has also been found to play a role in the production of pollen.

L – proline has a melting point of 228°C and a boiling point of 253°C whilst it rotates plane polarised light by -84°.

With a **formula mass** (M_r) of 115.13 g mol^{-1}, L – proline has the **elemental composition**: C: 52.12 %, H: 7.90 %, N: 12.17 %, O: 27.79 % so both its empirical and molecular formulas are $C_5H_9NO_2$

L – Proline

Infrared Spectrum

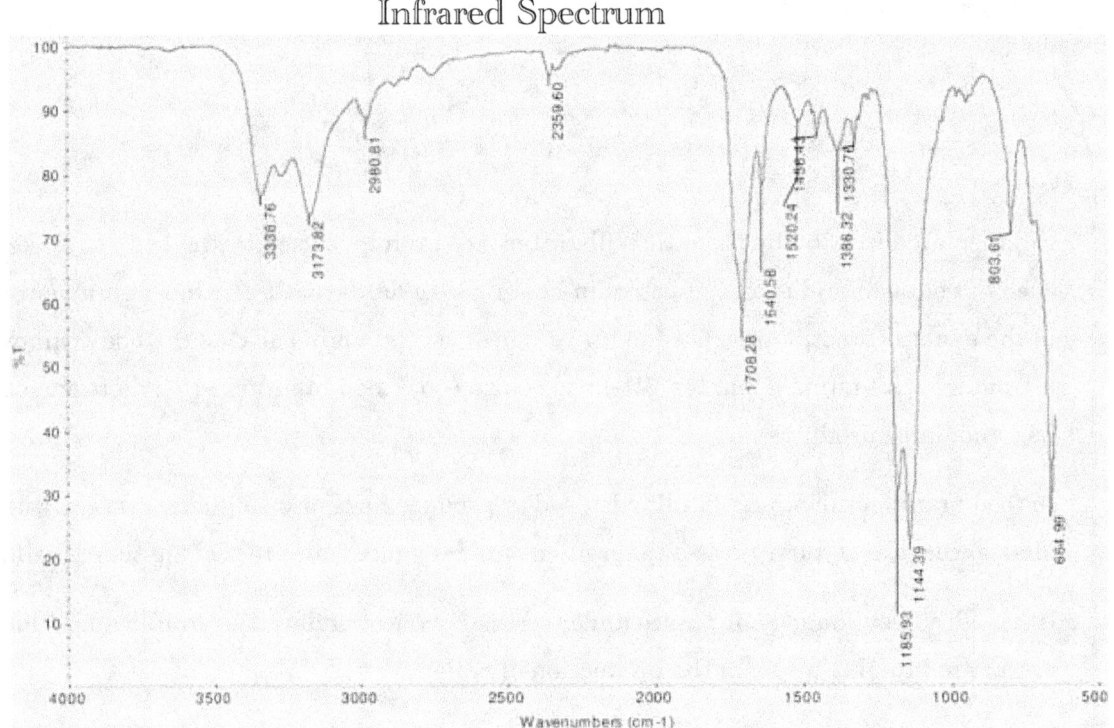

Observations

(√ / X)	Wavenumber range (cm⁻¹)	Wavenumber (cm⁻¹)	Assignment
√	3200 – 3700	3339	O – H
√	3200 – 3600	3250	N – H
√	3000 – 3300	3174	C – H (aromatic)
√	2500 – 3000	2980	C – H (aliphatic)
√	2200 – 2500	2360	C ≡ N
?	1700 – 1800	1708	C = O
?	1700 – 1800	1708	C = N
X	1600 – 1700		C = C (aliphatic)
X	1585 – 1600		C – C (aromatic)
X	1450 – 1600		C – C (aromatic)
√	1000 – 1300	1144	C – O
X	700 – 1000		C – X (X = Cl, Br or I)

Conclusions

- The peaks assigned to the O – H and N – H are tentative and, since the wavenumber ranges overlap, they can be the other way around.

- There is a small peak in the aromatic C – H region but this is unusually small for such a bond. Also importantly, the molecule contains only five carbon atoms so the likelihood of it being aromatic is extremely low.

- The molecule appears to contain a carboxylic acid group due to the C = O stretch and it is possible that there is also a C ≡ N bond but the peak at 2360 cm⁻¹ is very weak for a polar bond.

Mass Spectrum

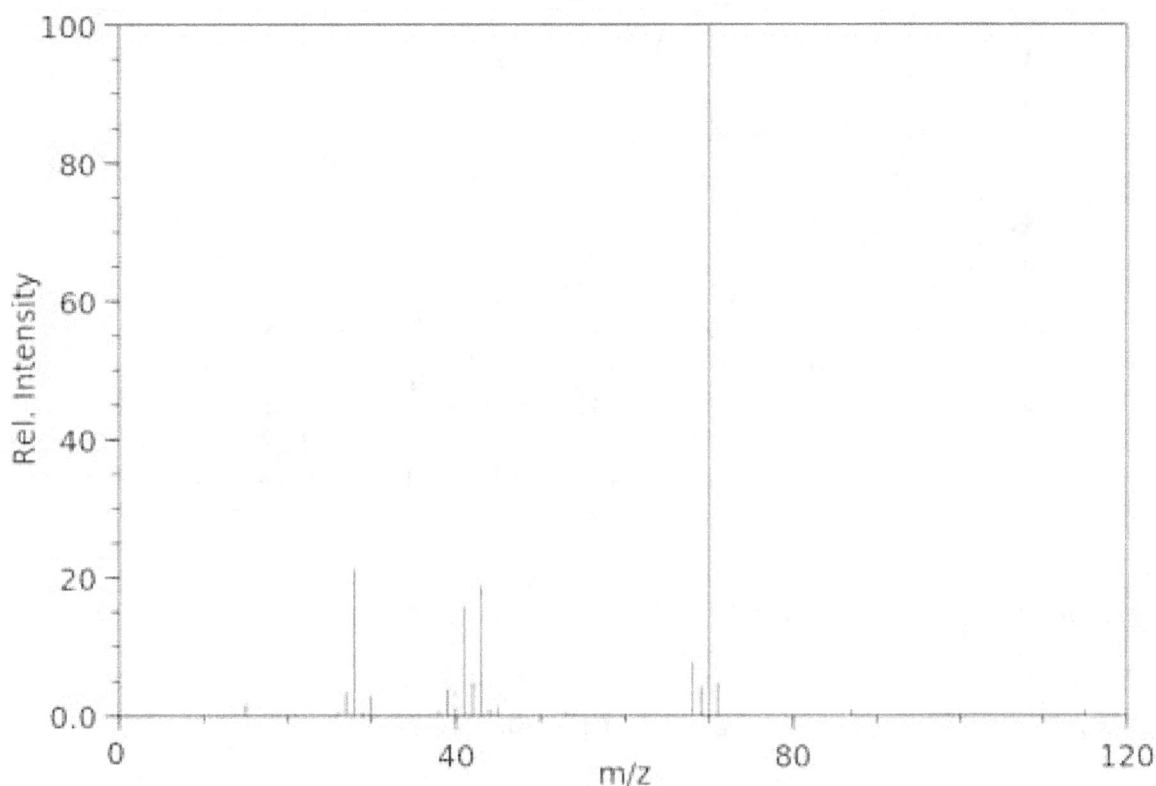

Observations

Charged fragments (m/z)	Assignment	Charged fragments (m/z)	Assignment
Molecular ion: 115	$[C_5H_9NO_2]^+$	Base peak: 70	$[C_5H_{10}]^+$

96	$[C_4H_2NO_2]^+$	43	$[C_3H_7]^+$
87	$[C_4H_7O_2]^+$	28	$[C_2H_4]^+$

Conclusions

There is no evidence for the presence of an aromatic ring and the peaks recorded imply the presence of a four or five carbon chain or a smaller, substituted chain.

There are a large number of isomers of formula $C_5H_9NO_2$ We can dismiss most of the isomers shown above since the infrared spectrum demonstrates the absence of both a C = C bond and a C = N bond and there is also only very weak evidence for a C ≡ N bond. There are also a number of possible, non-aromatic, cyclic structures. In total there are 347 possible isomers of formula $C_5H_9NO_2$ and we can establish the correct structure by examining the ¹H and ¹³C nmr spectra which is our next task.

NMR Spectra

We can identify the correct structure by examination of the ¹H and ¹³C nmr spectra. On many occasions we have predicted the spectra before examining the actual spectrum. In this case there are too many possible isomers to make that remotely feasible so we will jump into examining the nmr spectra directly, starting with the ¹³C nmr spectrum.

The ¹H and ¹³C nmr spectra are displayed below:

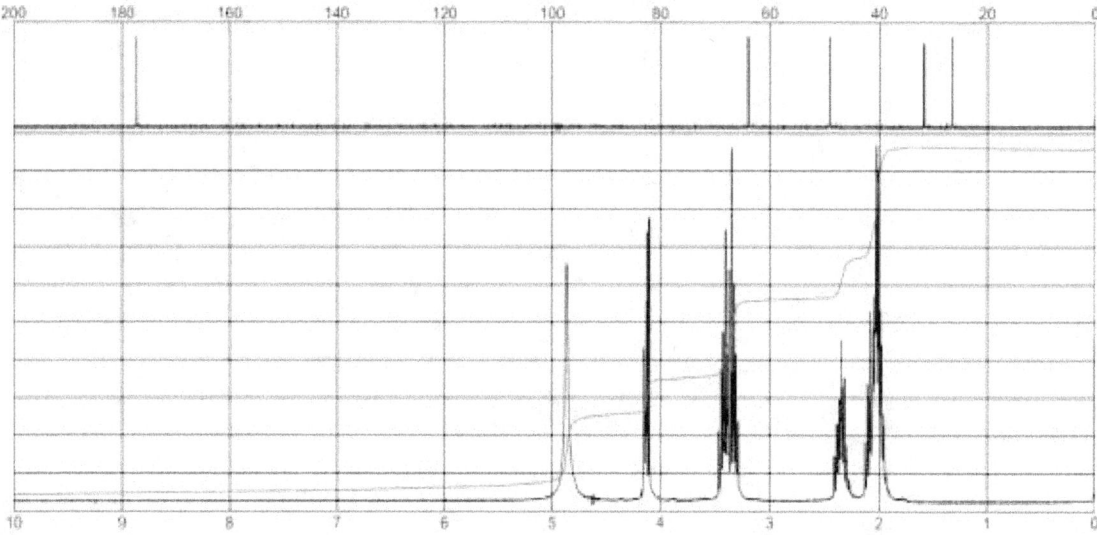

We will consider the ¹³C nmr spectrum first.

¹³C NMR Spectrum

It is clear from the spectrum that we can make the following assignments:

Chemical shift δ (ppm)	Integral	Assignment(s)
178	1	C = O
63	1	C – N
49	1	C – N
32	1	C – C
27	1	C – C

The first assignment is easy: the molecule must contain a carboxylic acid, –CO₂H, group which must be a terminal group – carboxylic functional groups are always terminal. We can draw a number of possible isomers but none of these are possible as none of them contain nine hydrogen atoms.

This means that the molecule must be cyclic and the ring could contain three or four carbon atoms perhaps with a nitrogen in the ring. A three – carbon ring, however, is not feasible as the remaining carbon atom would form a methyl, –CH₃, group and the number of hydrogen atoms would, again, not add up to nine.

This also means that four of the carbon atoms will form a ring but the intriguing assignments are the two C – N peaks which mean that the molecule must contain a nitrogen atom in the ring. The fifth carbon atom will form the –CO₂H group which must be a substituent and this means that we have the following candidates:

We can now predict the ¹H nmr spectrum and so the structures are repeated below but with the hydrogen atoms labelled alphabetically.

I   II   III

We can now predict the ^{1}H nmr spectrum for each of the three candidate molecules.

^{1}H NMR Spectrum Predictions

In all three cases there will be a singlet, of integral one, assignable to the –NH hydrogen atom. Similarly the hydroxyl, –OH, hydrogen if it appears will appear as, a singlet of integral one, anywhere in the same range so we will concentrate on the other hydrogen atoms.

Candidate I

* H_b will produce a doublet of triplets, of integral one, due to splitting by the two H_c hydrogen atoms (which are not chemically and magnetically equivalent) in the range δ 3 – 4.5 ppm due to its proximity to the nitrogen atom;

* The signal due to H_c will be a doublet of triplets, of integral two, due to splitting into a triplet by the two H_d atoms and then this triplet will be split into a doublet of triplets by H_b. This will also appear somewhere in the δ 0.5 – 2 ppm region;

* H_d will cause a triplet of triplets, of integral two due to splitting by the two H_c atoms, producing a triplet which will itself be split into a triplet of triplets by the two H_e atoms again in the δ 0.5 – 2 ppm region;

* The signal assignable to H_e will be a triplet, of integral two, due to splitting by the two H_d hydrogen atoms but this will appear somewhere between δ 3 ppm and δ 4.5 ppm.

Candidate II

* H_h will produce a triplet of triplets, of integral one, due to splitting by the two H_i hydrogen atoms and then by the two H_l atoms in the range δ 3 – 4.5 ppm due to its proximity to the nitrogen atom;

* The signal due to H_i will be a doublet of triplets, of integral two, due to splitting into a triplet by the two H_j atoms and then this triplet will be split into a doublet of triplets by H_b. This will appear somewhere in the δ 0.5 – 2 ppm region;

- H_j will cause a triplet of integral two due to splitting by the two H_i atoms, producing a triplet which appear in the δ 3 – 4.5 ppm region;

- The signal assignable to H_l will be a doublet, of integral two, due to splitting by the H_h hydrogen and will appear somewhere between δ 0.5 ppm and δ 2 ppm.

Candidate III

- H_n will produce a triplet of triplets, of integral one, due to splitting by the two H_o hydrogen atoms and then by the two H_r atoms in the range δ 3 – 4.5 ppm due to its proximity to the nitrogen atom;

- The signal due to H_o will be a doublet, of integral two, due to splitting into a triplet by the two H_n atoms and it will somewhere in the δ 3 – 4.5 ppm region;

- H_q will cause a triplet of integral two due to splitting by the two H_r atoms, producing a triplet which appear in the δ 3 – 4.5 ppm region;

- The signal assignable to H_r will be a triplet, of integral two, due to splitting by the H_q hydrogen atoms and will appear somewhere between δ 0.5 ppm and δ 2 ppm.

We can stop there as only Candidate I produces a doublet of triplets, of integral one, in the region δ 3 – 4.5 ppm so that must be the correct structure.

Conclusions

Structure:

Systematic name: L – proline and pyrrolidine-2-carboxylic acid.

Chapter XIV

L – Tryptophan

First isolated in 1901 from hydrolysed casein obtained from milk by Frederick Gowland Hopkins (Nobel Laureate in Medicine or Physiology (1929) for his discovery of vitamins), L – tryptophan is an essential amino acid.

As well as used in the synthesis of proteins, L – tryptophan is also converted by the body into serotonin which transmits signals between nerve cells. Low levels of serotonin have been linked to depression. L – tryptophan is also a precuror to melatonin and Vitamin B3.

L – tryptophan has also been claimed to be useful for holistic treatment of pre-menstrual stress and insomnia but evidence for its efficacy in these matters is limited.

L – tryptophan is found in a wide range of foods including dairy products, red meat, chicken and turkey as well as chocolate, dates and seeds.

It has a melting point of 290°C, a boiling point of 450°C and rotates plane polarised light by -30°.

L – tryptophan has a **formula mass** (M_r) of 204.23 g mol^{-1} and **elemental composition** of C: 64.63 %, H: 5.93 %, N: 13.72 %, O: 15.67 % meaning that its empirical and molecular formulas are $C_{11}H_{12}N_2O_2$.

L - Tryptophan

Infrared Spectrum

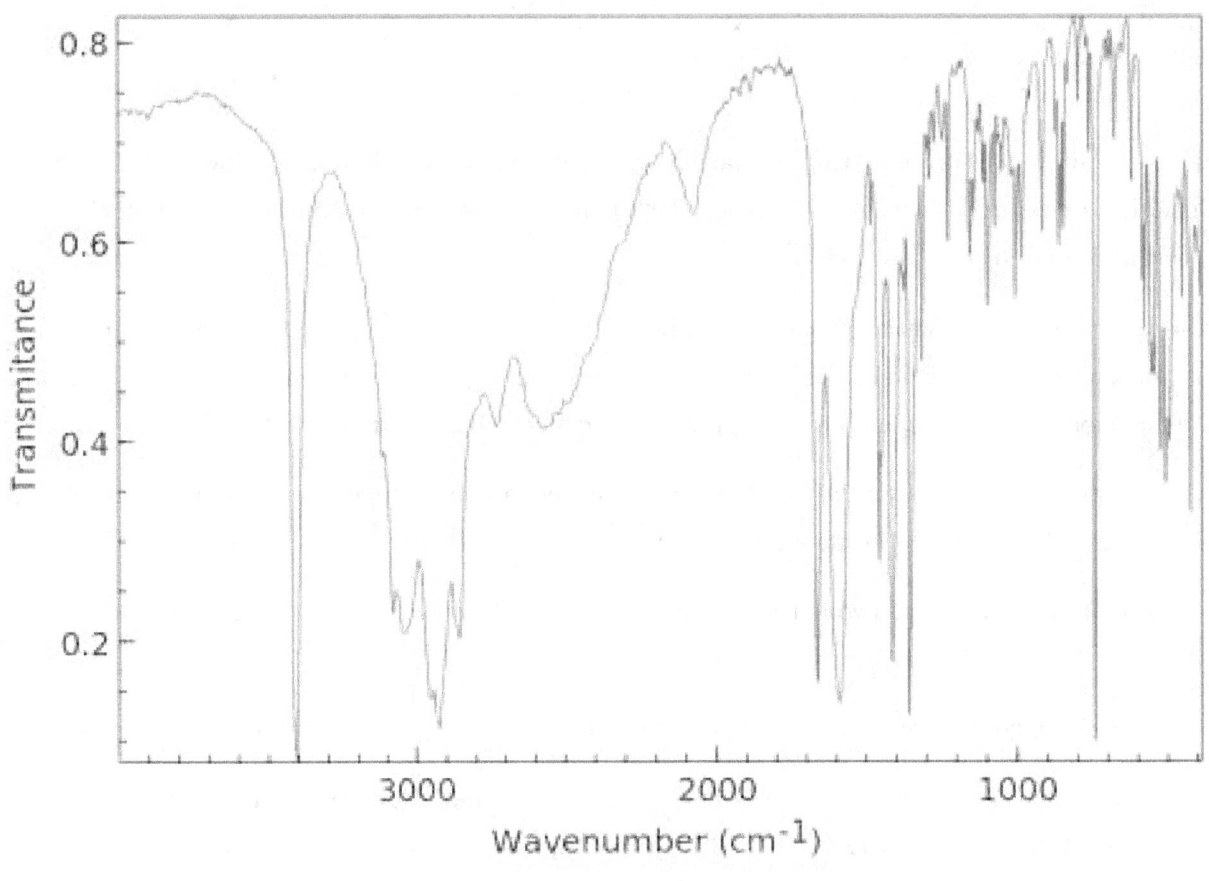

Observations

(√ / X)	Wavenumber range (cm⁻¹)	Wavenumber (cm⁻¹)	Assignment
?	3200 - 3700	3400	**O – H**
?	3200 - 3600	3400	**N – H**
√	3000 – 3300	3100, 3050	**C – H (aromatic)**
√	2500 – 3000	2950, 2920, 2850	**C – H (aliphatic)**
X	2200 – 2500		**C ≡ N**
?	1700 – 1800	1690	**C = O**
?	1700 – 1800	1690	**C = N**
?	1600 – 1700	1690	**C = C (aliphatic)**
√	1585 – 1600	1590	**C – C (aromatic)**
√	1450 – 1600	1450	**C – C (aromatic)**
√	1000 – 1300	1380	**C – O**
X	700 – 1000		**C – X (X = Cl, Br or I)**

Conclusions

This is an usual spectrum as it has a very sharp peak at 3400 cm⁻¹ which can be assigned to either O – H or N – H or both. The molecule is clearly aromatic and also has aliphatic C – H bonds and also possesses a carboxylic acid group.

Mass Spectrum

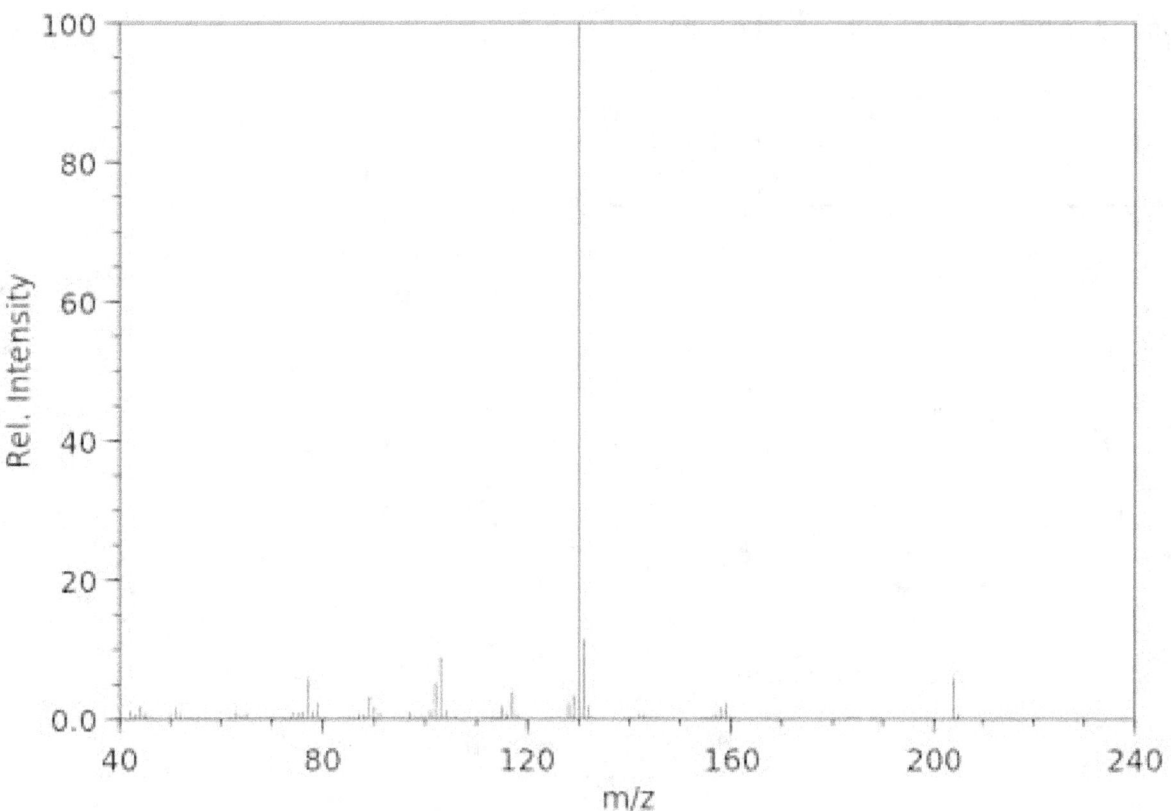

Observations

Charged fragments (m/z)	Assignment	Charged fragments (m/z)	Assignment
Molecular ion: 204	$[C_{11}H_{12}N_2O_2]^+$	**Base peak:** 130	$[C_8H_6O]^+$
159	$[C_{11}H_{11}O]^+$	103	$[C_8H_7]^+$
117	$[C_9H_9]^+$	76	$[C_6H_4]^+$

Conclusions

This spectrum contains a multitude of peaks which cannot be assigned but the most important peak is that at m/z = 76 which indicates the presence of a benzene ring with two substituents. Since the molecule contains a carboxylic acid, $-CO_2H$, group and a benzene ring with two substituents we can draft a basic structure as shown below where the squiggles indicate the, as yet, undetermined substituents.

These are the classic ortho-, meta- and para substituted benzene rings:

and leave us to account for the remainder of the formula $C_4H_7N_2$ which we can achieve by examination of the ^1H and ^{13}C nmr spectra.

NMR Spectra

The ^1H and ^{13}C nmr spectra are displayed below:

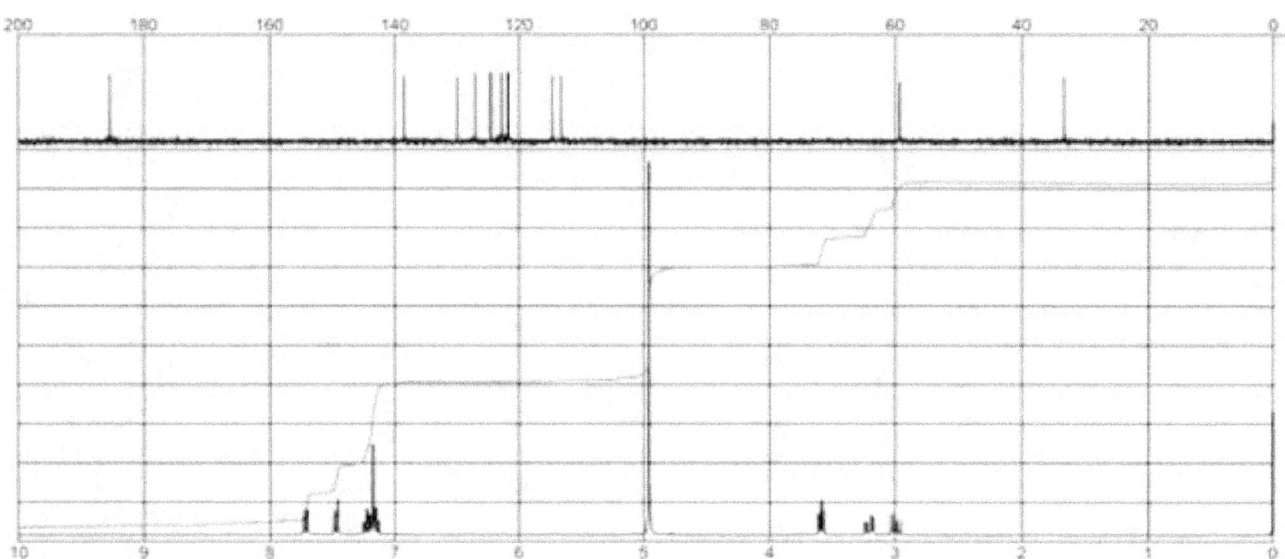

We will consider the ^{13}C nmr spectrum first.

^{13}C NMR Spectrum

Observations

We can make the following assignments:

Chemical shift δ (ppm)	Assignment(s)	Integral
185	C = O	1
138	Aromatic	1
130	Aromatic	1
127	Aromatic	1
125	Aromatic	1
123	Aromatic	1
122	Aromatic	1
117	C = C	1
115	C = C	1
59	C – N	1
34	C – C	1

Conclusions

▪ The peak at δ 185 ppm clearly indicates that the molecule does contain a carboxylic acid, –CO₂H, functional group.

▪ This must lead us to conclude that the signal at δ 59 ppm is due to a C – N bond. The C = O and C = N ranges overlap but there are only two oxygen atoms in the molecule, both bonded to the same carbon atom and so this peak cannot be assigned to a C – O carbon atom as that carbon atom has been accounted for already.

❋ There are two signals assignable to C = C carbon atoms. These atoms must form either side of the same bond but since they produce signals at different, but very similar, chemical shifts the bond must be asymmetric i.e. have different substituents on either side of the bond. The simplest structure to consider, and the simpler the better – at least initially is that the alkene, C = C bond is attached to the benzene ring.

❋ But we can go further by examining the aromatic carbon signals. There are six plus two other peaks in this range so indicating that all the carbon atoms are chemically and magnetically *non-equivalent* and this means that we can eliminate the para-substituted isomer as that molecule would contain two pairs of chemically and magnetically equivalent carbon atoms.

❋ One of the aromatic peaks, at δ 138 ppm, and significantly away from the other aromatic and alkene carbon signals indicates that this carbon atom is significantly more deshielded. This implies the attachment of an electronegative atom which, in this case must be nitrogen. Since there is only one C – N signal, the other nitrogen atom must be part of a substituent, perhaps an amino, –NH$_2$, group.

❋ We, therefore, have the following possibilities:

and the squiggles must represent C$_4$H$_5$N as everything else is accounted for.

❋ The squiggle must also include a C = C bond so we can redraw these possible structures as

so that leaves us with C$_2$H$_4$N to account for but we cannot construct a linear chain from this fragment which implies that the substituent –NH$_2$ is not correct and can only be –NH and then that can only occur if the nitrogen atom in the aromatic NH substituent is bonded to a carbon atom.

This means that we can discard the meta – isomer as the two substituents must be on adjacent carbon atoms and this leaves us with:

It is not possible for the other nitrogen atom to be adjacent to the other nitrogen atom and so there must be another carbon atom.

❉ This gives us the following possibilities:

where the squiggle in the ring is either a single C – C bond or a C = C bond. There are insufficient hydrogen atoms to allow it to be a single bond and so we can now draw it as one of the structures below:

The squiggle represents C_2H_5N and this leaves us with four possibilities:

❉ We can predict the spectra for each of these candidates using alphabetic labelling for each of the hydrogen atoms and will do so below but it is worth examining the benzene ring first.

In all four cases the ^1H nmr peaks will be the same

- ❉ H_a will produce a doublet of integral one due to splitting by H_b;

- ❉ The signal due to H_b will be a doublet of doublets due to sequential splitting by H_a and H_c;

- ❉ Likewise H_c will be a doublet of doublets, of integral one, due to sequential splitting by H_b and H_d;

- ❉ As with H_a, the hydrogen atom H_d will produce a doublet of integral one due to splitting by H_c.

This is exactly what is observed and the aromatic ring does not need further consideration when predicting the spectrum for each of the four candidates however the base structure is interesting as it basically a substituted indole ring.

Tryptophan can be considered to be a substituted with this basic structure:

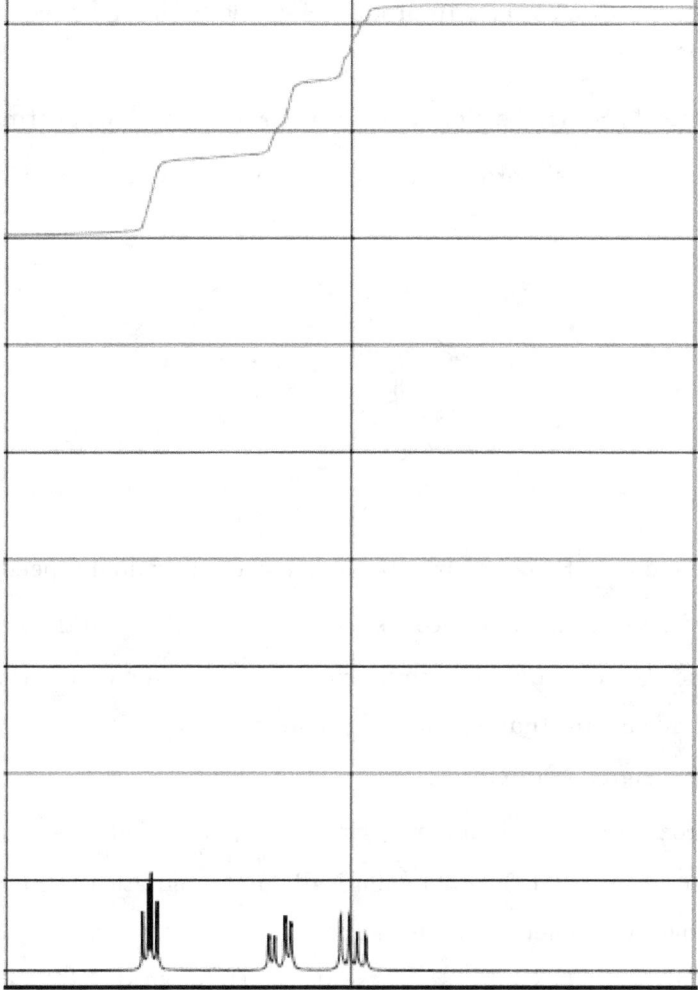

The base structure is interesting as it is also the basis for serotonin which has multiple uses including regulating bowel behaviour and also peoples' moods etc;

There is, however, a singlet of integral one also the δ 7 – 8 ppm region which must be due to the alkene hydrogen atom. This peak is at δ 7.4 ppm and alkene hydrogen atoms can appear anywhere between δ 5 ppm and δ 8 ppm. The peak at δ 4.85 ppm is due to water.

The expanded ¹H nmr spectrum (δ 4 -2 ppm) is shown below.

We can, however, learn much from predicting the spectra of the four candidate molecules which is our next task. Note that the structures are individually, alphabetically labelled.

★ Candidate I

- ★ H_e will not be deshielded by proximity to an electronegative atom and so will be found in the lower half of the δ 5 – 8 ppm range;

- ★ H_f can appear anywhere in the δ 0 – 12 ppm region and will be a singlet of integral one;

- ★ H_g will be responsible for a triplet of integral one due to splitting by the the two H_i atoms. The two H_i atoms will be chemically and magnetically equivalent as the carbon – carbon bond can rotate freely.

- ★ The two H_h hydrogen atoms will produce a singlet of integral two anywhere in δ 0 – 12 ppm range.

- ★ The signal due to H_i will be a doublet of integral two due to splitting by H_g.

- ★ If it appears then H_j will produce a singlet of integral one.

★ Candidate II

- ★ The signals due to H_a – H_d have already been discussed so do not need further analysis.

- ★ As with Candidate I, H_e will produce a singlet of integral one in the lower half of the δ 5 – 8 ppm region since there is in its proximity to deshield it. Remember that deshielding results in a higher chemical shift than shielded equivalents.

- ★ H_f will produce a signal of integral one;

- ★ Due to their proximity to the indole ring, H_g and $H_{g'}$ are labelled separately as a consequence of their location means that they not chemically and magnetically equivalent. Each will produce a doublet of doublets of integral one since:

- ★ The signal due to H_g will be split into a doublet by $H_{g'}$ and this doublet will be split into a doublet by H_h whilst;

- ★ That due to $H_{g'}$ will split into a doublet by H_g and this doublet will be split into a doublet by H_h;

- ★ H_h will produce a doublet of doublets of integral one due to splitting by H_k and $H_{k'}$.

- ★ The signal due to H_i will be a singlet of integral two. There will be no splitting since nitrogen does not couple with adjacent hydrogen atoms.

Candidate III

- As with the other candidates H_e will produce a singlet of integral one.

- H_f will cause a singlet of integral one. It will be deshielded by its proximity to the electronegative nitrogen atom and so the singlet will appear in the upper half of the δ 5 ppm – 8 ppm region.

- H_g will produce a triplet, of integral one, due to splitting by the two H_j atoms. In this case, the latter two atoms, H_j, will be chemically and magnetically equivalent as there is free rotation around the C – C bond.

- H_h will, as usual, be responsible for a singlet of integral two.

Candidate IV

- H_f will produce a singlet, of integral one, and being deshielded by the adjacent nitrogen atom will appear in the upper half of the δ 5 – 8 ppm range;

- H_g and $H_{g'}$ are labelled separately due to their proximity to a planar ring;

  - H_g will produce a singlet which is split into a doublet by $H_{g'}$ which is then split into a doublet of doublets, of integral one, by H_h;

    - $H_{g'}$ will also produce a doublet of doublets , of integral one, due to sequential splitting by H_g and H_h. Although H_g and $H_{g'}$ are not chemically and magnetically equivalent there is, otherwise, very little difference between them and they will appear in almost exactly the same position. It is difficult to predict exactly where this will appear because the atoms will be deshielded by H_h because that has been deshielded by the nitrogen atom and this is one occasion where we will rely on the multiplicity and the integral rather than the chemical shift;

  - The single H_h atom will produce a doublet of doublets, of integral one, due to successive splitting by H_g and $H_{g'}$ in the δ 3 – 5 ppm region;

  - H_i will produce a singlet, of integral two, somewhere between δ 3 ppm and δ 5 ppm.

It is clear that the only plausible structure is candidate IV.

Conclusions

Structure:

Systematic name: Tryptophan and (2S)-2-amino-3-(1H-indol-3-yl)propanoic acid.

Chapter XV

L – Tyrosine

First discovered, in 1846, by Justus von Liebig from casein found in cheese, L – tyrosine is a non-essential amino acid. The name comes from the Greek *tyrós*, meaning cheese and performs numerous roles in the body as well as being involved in photosynthesis in plants.

L – tyrosine is biosynthesised from L – phenylalanine (Chapter XII) but it is also found in a wide range of foods including white meat, dairy products, seeds, avocados and bananas.

It is active in the brain and the whole body and is a precursor to adrenaline, dopamine, thyroid hormones and melanin. Widely available as dietary supplement, there have been claims that it improves alertness, attention and focus but there is little clinical evidence for these claims. There is, however, evidence that it can be used to treat phenylketonuria which is an inherited disorder and it is also used for alcohol and cocaine dependence. It has been claimed to be involved the body's circadian rhythm and might improve mental health.

L – tyrosine has a melting point of 300°C, a boiling point of ~390°C and it rotates plane polarised light by -11°.

With a **formula mass** (M_r) of 181.19 g mol^{-1} , L – tyrosine has the **elemental composition**: C: 59.61 %, H: 6.13 %, N: 7.73 %, O: 26.49 % and this means that it has the empirical and molecular formulas of $C_9H_{11}NO_3$.

L – Tyrosine

Infrared Spectrum

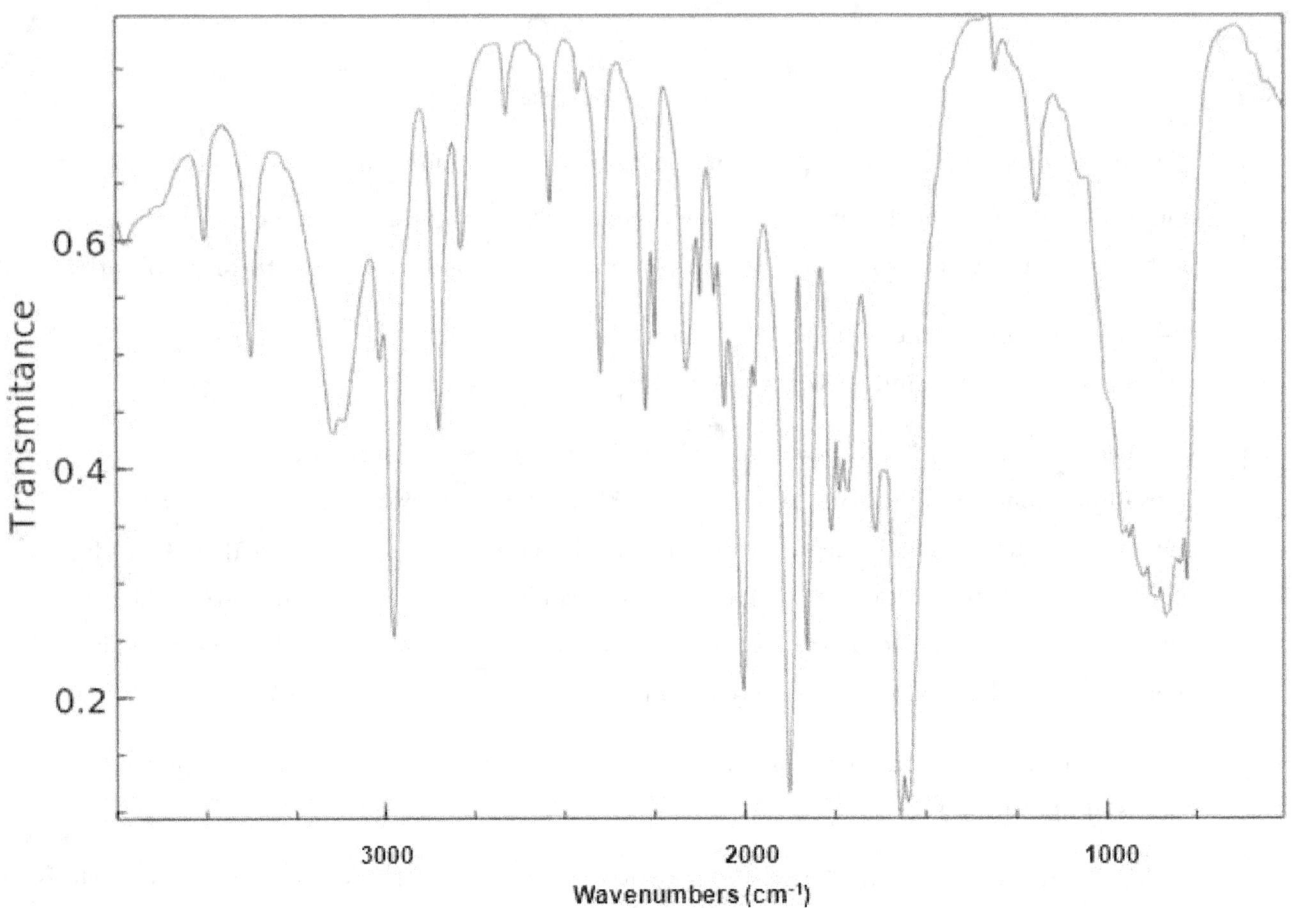

Observations

(√ / X)	Wavenumber range (cm⁻¹)	Wavenumber (cm⁻¹)	Assignment
√	3200 - 3700	3550	**O – H**
√	3200 - 3600	3400	**N – H**
√	3000 – 3300	3100	**C – H (aromatic)**
√	2500 – 3000	2995, 2800	**C – H (aliphatic)**
X	2200 – 2500		**C ≡ N**
√	1700 – 1800	1790	**C = O**
X	1600 – 1700		**C = C (aliphatic)**
√	1585 – 1600	1580	**C – C (aromatic)**
X	1450 – 1600		**C – C (aromatic)**
√	1000 – 1300	1280	**C – O**
X	700 – 1000		**C – X (X = Cl, Br or I)**

Conclusions

It appears that this molecule:-

- Is aromatic with an aliphatic carbon – carbon chain;

- Contains both O – H and N – H groups;

- Has a terminal carboxylic acid, $-CO_2H$, group.

We will learn more from the mass spectrum and the ^1H and ^{13}C nmr spectra.

L – Tyrosine

Mass Spectrum

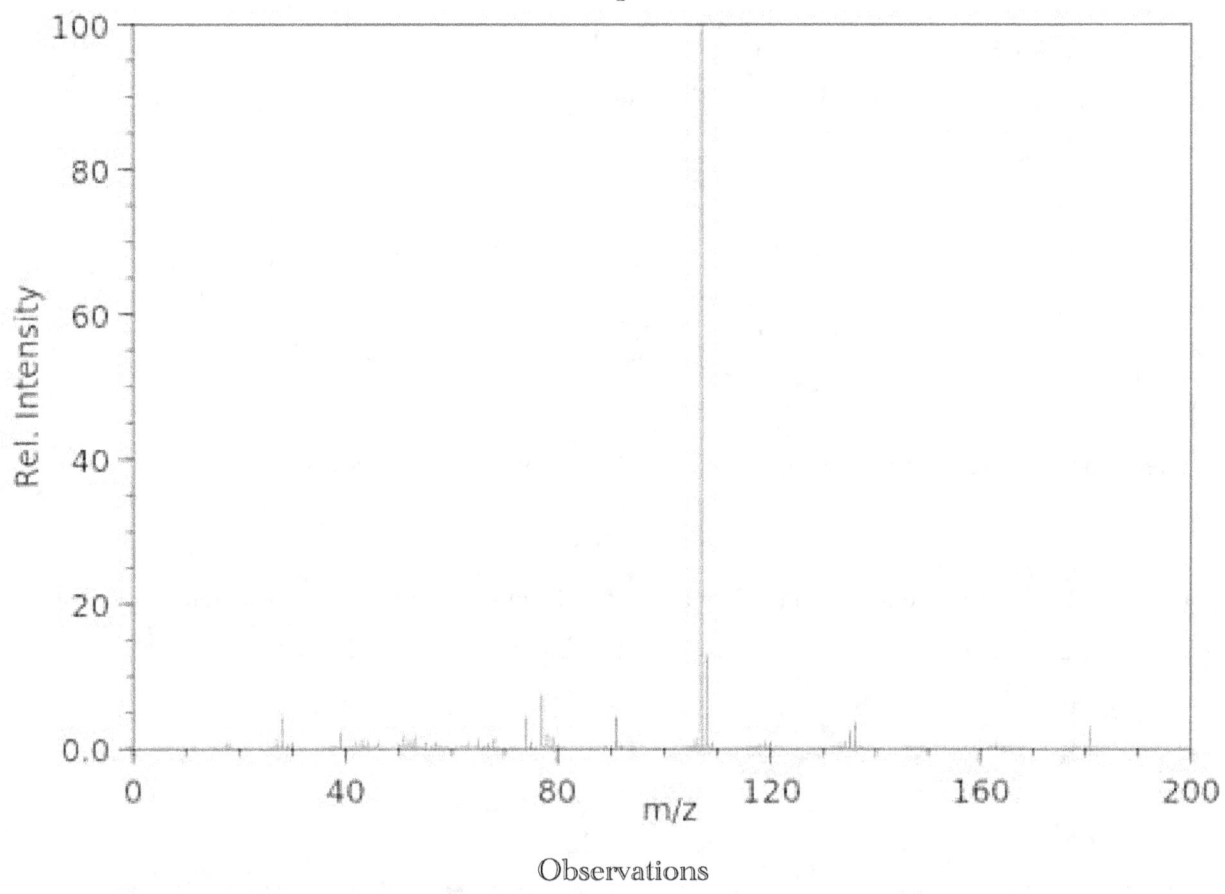

Observations

Charged fragments (m/z)	Assignment	Charged fragments (m/z)	Assignment
Molecular ion: 181	$[C_9H_{11}NO_3]^+$	**Base peak:** 107	$[C_7H_7O]^+$

136	$[C_8H_{10}NO]^+$	76	$[C_6H_4]^+$
91	$[C_7H_7]^+$	45	$[CO_2H]^+$

Conclusions

Whilst the molecular ion is simply the ionised molecule, the peak at:-

- m/z = 76 indicates the existence of a di-substituted benzene ring;
- m/z = 107 suggests the existence of a $C_6H_4OH\text{-}CH_2$ group;
- m/s = 91 implies the existence of a $C_6H_4H\text{-}CH_2$ group;
- m/z = 45 is due to a $[CO_2H]^+$ group.

This means that molecule probably comprises a substituted phenol ring. The second substituent comprises a carbon chain terminating in a carboxylic acid, $-CO_2H$ group, giving us three possible candidates where the squiggle represents the remainder of the carbon chain:

The squiggle represents $C_2H_5NH_2$ and this means that there are six possible isomers as shown below:

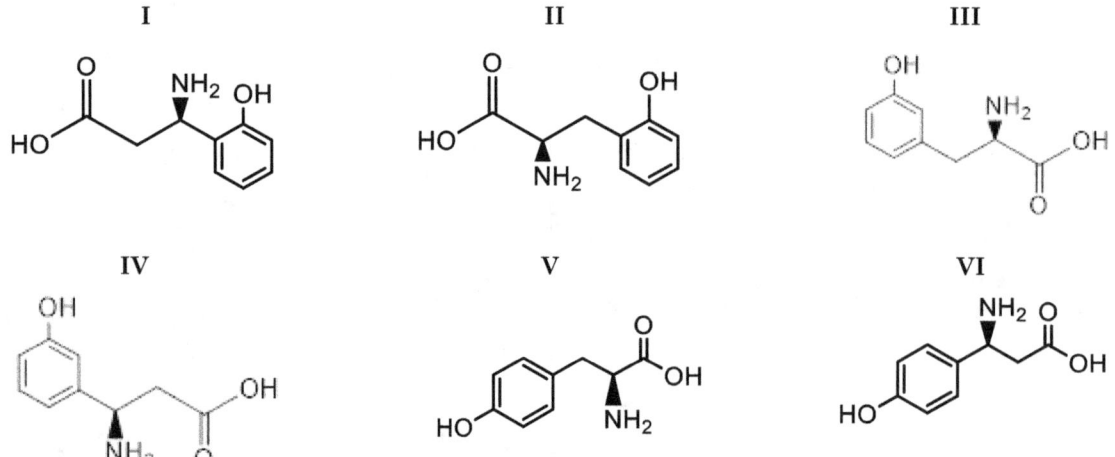

We can determine the correct structure by examining the ¹H and ¹³C nmr spectra.

NMR Spectra

The ¹H and ¹³C nmr spectra are displayed below and we will consider the ¹³C nmr spectrum first.

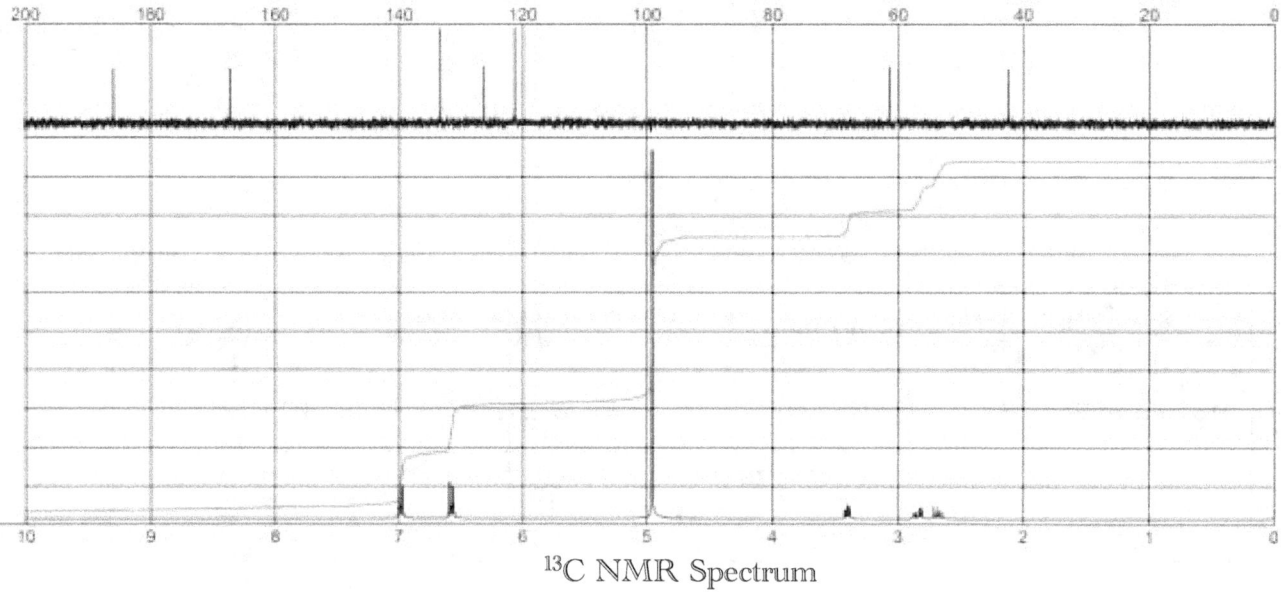

¹³C NMR Spectrum

We can make the following assignments,

▪ The peak at δ 175 ppm must be due to a carbonyl, C=O, carbon atom;

▪ The four peaks in the δ 110 ppm – δ 160 ppm region can be assigned to aromatic carbon atoms.
  The peak at δ 158 ppm, indicating significant deshielding, implies that this carbon atom is bonded
  to an electronegative atom and this would be either a nitrogen or, more likely, an oxygen atom
  since the mass spectrum indicates that the molecule is a substituted phenol.

▪ The two pairs of peaks of integral two indicates that there are two pairs of chemically and magnetically
  equivalent aromatic carbon atoms and the ring must have two substituents. In addition this also means
  that the ring must have the two substituents on opposite sides of the ring i.e. in the para- position;

▪ The peak at δ 58 ppm is due to a C – N bond since the two oxygen atoms have already been accommodated;

▪ The peak at δ 37 ppm must be due to a C – C bond which must be on the chain.

This means we can discard four of the possible candidates leaving just isomers V and VI which can be
distinguished by the ¹H nmr spectrum.

L – Tyrosine

^{1}H NMR Spectrum

Before we examine the actual spectrum we will predict the spectra that will be produced by the two candidate molecules which is a useful exercise as it deepens our understanding of nmr spectroscopy.

Candidate V

◾ We may or may not observe peaks due to the hydroxyl hydrogen atoms, H_a and H_i and the amino hydrogen atoms, H_h, as they can be labile. If we do observe them then they can appear anywhere in the $\delta 0 - 12$ ppm region;

◾ H_b and H_d are mutually chemically and magnetically equivalent;

   ◾ H_b will produce a doublet, of integral one, due to splitting by H_c.

   ◾ Similarly H_c will produce a doublet, of integral one, due to splitting by H_b;

◾ H_c and H_e are also mutually chemically and magnetically equivalent.

   ◾ H_d will produce a doublet, of integral one, due to splitting by H_e. Likewise, H_e will produce a doublet of integral one, due to splitting by H_d;

This means that, in both cases, we should observe two pairs of doublets, all of integral one, in the aromatic, $\delta 6 - 8$ ppm region;

◾ Because of their proximity to H_g, the stereochemistry of the *two* H_f atoms will produce two doublets, of integral two. They will, however, appear close together ;

◾ H_g will produce a triplet, of integral one, due to splitting by the two H_f atoms.

It will appear at the upper end of the range of the $\delta 3 - 5$ ppm but might be even higher due to its proximity to the two electronegative atoms and to the nitrogen atom;

◾ The two H_h atoms will produce a singlet of integral two.

Candidate VI

◾ As with candidate V, the aromatic hydrogen atoms, H_k, H_l, H_m and H_n will produce two doublets of integral two in the $\delta 7 - 8$ ppm;

◾ H_o will produce a triplet, of integral one, due to splitting by H_p;

◾ The two amino hydrogen atoms, H_q, will produce a singlet of integral two.

If we examine the expanded ^{1}H nmr spectrum we can immediately see that it matches the predictions for candidate V and we can make some definite assignments.

The expanded ^{1}H nmr spectrum, which is shown below, clearly matches candidate V and the assignments are shown below the spectrum. The singlet at δ 6.3 ppm is due to water. Whether or not it is from contamination of the solvent or the sample is uncertain but analysis of contaminated and fresh samples will show a variable integral at the chemical shift.

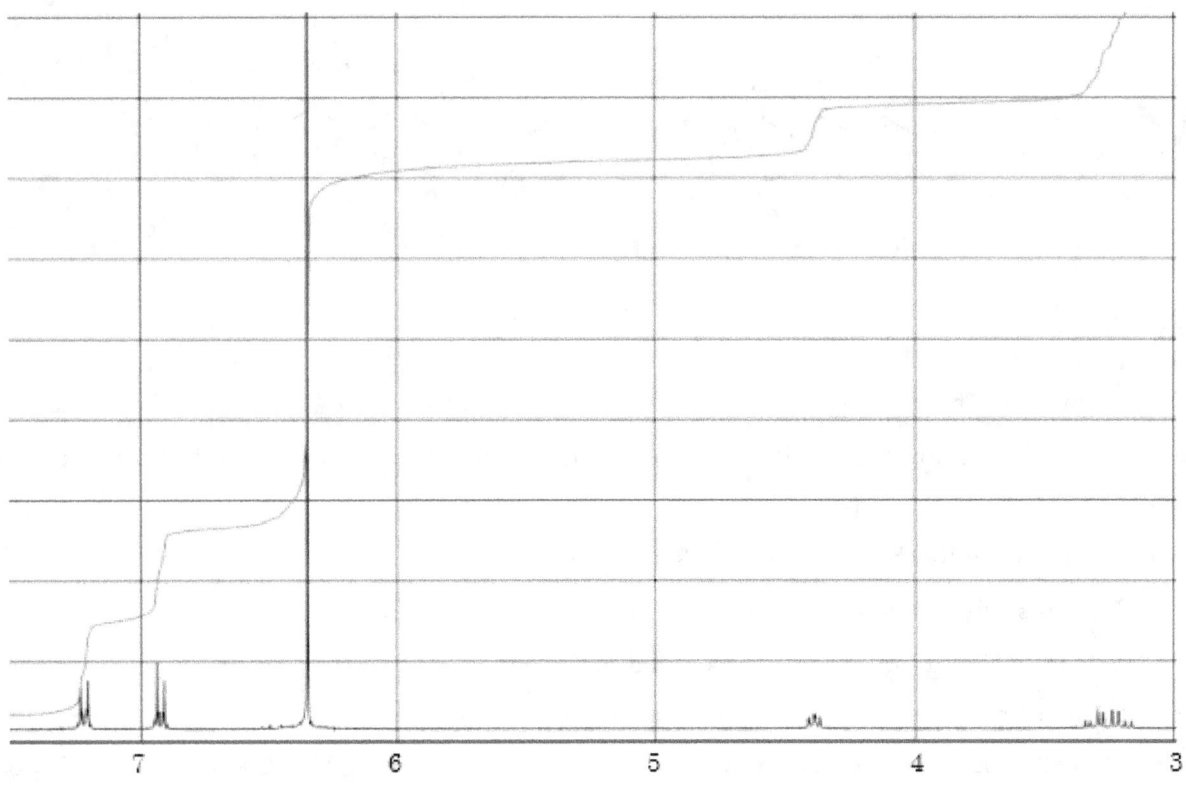

We can make the following assignments which can only match the structure of candidate V.

Chemical shift δ (ppm)	Integral	Multiplicity	Assignment
7.25	1	Doublet of doublets	Aromatic hydrogen
7.20	1	Doublet of doublets	Aromatic hydrogen
6.95	1	Doublet of doublets	Aromatic hydrogen
6.90	1	Doublet of doublets	Aromatic hydrogen
5.15	2	Singlet	H_h
4.35	2	Triplet	H_g
Centred on 3.25	1	Doublet of doublets	One of the two H_f atoms
Centred on 3.25	1	Doublet of doublets	One of the two H_f atoms

It is not possible to specifically assign the aromatic hydrogen atoms but the prediction of the aromatic ring as being para- isomer is clearly correct. Likewise it is not possible to say which doublet of doublets can be assigned to specific H_f atoms.

L – Tyrosine

Overall Conclusions

Structure:

Systematic name: L – tyrosine and L – 2 – amino – 3 – (4-hydroxyphenyl)propanoic acid.

Part III

Pen Portraits

Most science textbooks concentrate solely on the theory and practice of the subject but we believe that it is important to know something about those who discovered, isolated and characterised these molecules.

This, the final section of the book, provides very brief pen portraits of the scientists who made important discoveries about amino acids.

The discoveries are, in themselves, remarkable and the discoverers are of interest in their own right. Not all amino acids were discovered by chemists seeking to identify them. Adolphe Strecker, for example, synthesised L – alanine and it was only later that it was found to be a non – essential amino acid.

It is also interesting that many of the people discussed in this volume collaborated with, or studied under others from time to time. For example, Strecker had studied under Justus von Liebig who, amongst many other achievements, had isolated and characterised L – tyrosine.

It is also worthwhile remember that compared to today, analytical tools were extremely limited and often amount to little more than elemental composition, molecular weight (as it was termed at the time although now is called the formula mass) melting and boiling points and recognised chemical tests. The last is one reason why students are still trained in conducting qualitative tests.

In current times, we have an array of spectrometric and spectroscopic techniques as well as x-ray crystallography etc; Whilst, quite often, 19th century chemists were wrong it is quite extraordinary just how often they were correct.

The brief pen portraits are written in alphabetical order by surname.

Henri Braconnot

Henri Braconnot (29th May 1780 – 13th January 1855) was born in Commercy. At the age of thirteen he became an apprentice with an apothecary in Nancy.

Other than a couple of years in Paris, Braconnot spent his life in Nancy where he concentrated on botany and the chemistry of natural products.

A prolific writer of over one hundred papers, Braconnot explained the process of saponication (soap making). He is also largely credited with the discovery of pyrogallic acid (1,2,3 – trihydroxybenzene) although this has also been credited to the Swedish chemist Carl Wilhelm Scheele. This compound which led to the invention of photography and established the process of the conversion of any of wood, straw or cotton into glucose.

From starch and honey separately, he also established, isolated and characterised the amino acid, L – glycine.

Ferdinand Heinrich Edmund Drechsel

Leipzig – born, Heinrich Ferdinand Edmund Drechsel (3rd September 1843 – 22nd September 1897) studied at the University of Leipzig before joining the Carl Ludwig Institute of Physiology in Leipzig in 1872.

It was at Leipzig (1889) that Drechsel discovered the amino acid L – lysine by isolating it from casein and also demonstrated that proteins are the main source of urea found in urine whilst also working on the chemistry of melamines, carbamates and cyanamides.

In 1889, L – lysine was first isolated from milk the protein casein but he also isolated it from other dairy products, red meat, fish and lentils.

Drechsel is perhaps even better known for his invention (1875) of the laboratory wash bottle which is familiar to all chemistry students. the original bottles, known as Drechsel bottles were made of glass but are now usually made from polyethylene terephthalate (PET). This device became massively popular and Drechsel could have made a fortune if he had patented it.

Hermann Emil Louis Fischer

Born near Cologne, Hermann Emil Louis Fischer FRS FRSE FCS (9th October 1852 – 15th July 1919) is one of the most famous organic chemists and in recognition of his multiple achievements he was awarded the 1902 Nobel Prize in Chemistry. The actual Nobel Prize citation states that it was awarded *'in recognition of the extraordinary services he has rendered by his work on sugar and purine* syntheses.'*

Fischer is renowned , always known as Emil, developed the Fischer projection for drawing asymmetric molecules and he also devised the *'lock and key'* model of enzyme actions and isolated and characterised L – valine.

He studied, firstly, at the University of Bonn in 1871 before moving to the University of Strasbourg in 1872. Fischer studied under Adolf von Baeyer (Nobel Prize in Chemistry, 1905, for *'in recognition of his services in the advancement of organic chemistry and the chemical industry, through his work on organic dyes and hydroaromatic compounds.'*

Fischer was awarded his doctorate in 1874 for his work on phthaleins**.

After eight years at Strasbourg, he was appointed to the chair of chemistry at the University of Erlangen in 1882 before moving to the University of Würzburg in 1885. He then moved to the University of Berlin in 1892.

Fischer's interest were multiple and diverse and included preparation of derivatives of hydrazine, diazo compounds, magenta dyes, barbiturates and was the first person to synthesise caffeine and D-(+)-glucose.

Taking an interest in proteins determined that proteins are composed of amino acids and he defined the linkage as a *'peptide bond'* although the term *'amide bond'* is now preferred and he proved this by synthesising proteins comprising long chains of amino acids.

* *Purine* is a heterocyclic aromatic organic compound comprising two fused rings (pyrimidine and imidazole). Substituted rings give rise to a whole class of molecules known as purines. To date, they are believed to be the most widely naturally occurring nitrogen-containing heterocycles.

** *Phthaleins* are a subclass of triarylmethane compounds i.e. three aromatic substituents on a methane backbone. They are formed by the reaction of phthalic anhydride with any of a large number of substituted phenols. Used as dyes and as pH indicators the most famous is phenolphthalein which is pink in alkaline solutions and becomes colourless in acidic solutions.

Sir Frederick Gowland Hopkins OM PRS

Sir Frederick Gowland Hopkins (20th June 1861 – 16th May 1947) was awarded the Nobel Prize in Physiology or Medicine in 1929, which he shared with Christiaan Eijkman *'for his discovery of the growth-stimulating vitamins.'*

Born in Eastbourne he studied at Birkbeck College and then the medical school at Guy's Hospital. After graduating he spent four years (1894 – 1898) teaching at Guy's before moving to Emmanuel College, Cambridge to lecture in physiology. He discovered the amino acid, L – tryptophan, in 1901. Awarded a doctorate from the University of London in 1902 he was appointed as a reader at Trinity College, Cambridge and was appointed as the first professor of biochemistry at Cambridge.

Hopkins had a deep interest in energy sources for cells and established that cells obtain energy through a complex series of oxidations and reductions. He also showed that lactic acid plays a major role in muscle contraction.

Hopkins is best known for his seminal work on animal feed (1912) through which he demonstrated that a diet comprising all or combinations of proteins, fats, carbohydrates, minerals and water did not promote animal growth.

From these experiments he surmised that there must be other, as yet unknown, substances in animal feed that do promote growth. Since they had not been isolated he also hypothesised that they must be present in very small quantities and he termed them *'accessory food factors'*. After some had been discovered they were termed *'vital amines'* until it was discovered that not all are actually amines hence they are now known as *'vitamins'*.

During World War I, Hopkins studied the nutritional content of margarine and established that it was inferior to butter as industrially produced margarine does not contain vitamins A and D. Vitamin-enriched margarine was introduced in 1926 and is one of the first food products to have had food additives added in the manufacturing process.

Hopkins is also credited, in 1921, with discovering, isolating and characterising glutathione. Glutathione has been implicated in many diseases including the protection of cancer cells and a cause of dementia.

Beyond his academic, scientific discoveries, Hopkins was a quite extraordinary person. Before the Second World War he took an active part in helping several German Jewish scientists including Hans Krebs who became famous for the Krebs (citric acid) Cycle.

Despite criticism, he was a great supporter of women in science research which is especially significant since at the time female students could not graduate from Cambridge. A significant number of the women he employed as his assistants went on to make important discoveries.

Among them were Marjory Stephenson who wrote a seminal textbook *Bacterial Metabolism* (1930) and was one of the first two women elected to be a fellow of the Royal Society and Dorothy Needham.

Dr. Dorothy Needham (née Moyle) was a most remarkable woman. She attended Girton College, Cambridge where she became fascinated by chemistry after attending some of Hopkins' lectures and obtained her Ph.D. in 1930 under his mentorship. She married a notable biochemist, Dr. Joseph Needham, another of Hopkins' students who became the the first head of the Natural Sciences Section of UNESCO in Paris and it was him who insisted that Science should be included in its full title of *United Nations Educational, Scientific and Cultural Organization*.

Dorothy Needham was also fundamental in establishing two colleges in Cambridge, New Hall (1946), and Lucy Cavendish College (1962) to accommodate female researchers who had no college appointments.

They are also notable for being the first married couple to be, simultaneously, Fellows of The Royal Society.

Justus von Liebig

Darmstadt – born Justus Liebig (May 12th 1803 – April 18th 1873) made analytical chemistry accessible to a generations but is probably known for his Liebig condenser for collecting liquid fractions of a mixture of compounds.

Liebig studied at the University of Bonn and obtained his doctorate at the University of Erlangen and, in 1822, went to study in Paris where he worked in the laboratories of Joseph Louis Gay-Lussac, most famous for determining the empirical formula of water.

Appointed professor of chemistry at the University of Giessen in 1827, Liebig established a large laboratory and very soon attracted many students of analytical chemistry but also studied organic chemistry.

Together with Friedrich Wöhler, Liebig synthesised the first laboratory produced organic compound, uric acid, disproving the concept that organic compounds have a '*vitality*' which compounds of metals did not have. They also produced the first isomers of silver fulminate (AgCNO, and silver cyanate, AgOCN).

In 1832 Liebig founded and became the editor of the journal '*Annalen der Chemie*' which is still one of the most important and prestigious chemical journals.

In his later years, Liebig returned to studying nutritional chemistry and hypothesised about the importance of nitrogen, carbon dioxide and minerals in plant growth. He also devised a method of extracting juices from beef carcasses. The dried extract is still extremely popular as Oxo™ cubes.

From our perspective, however, Liebig's most important discovery, isolation and characterisation of the amino acid, L – tyrosine, which he extracted from casein from cheese in 1846.

John Howard Mueller

Massachusetts – born John Howard Mueller (June 13th 1891 – February 14th 1954) is famous for his discovery of the amino acid L – methionine in 1921.

A clergyman's son, Mueller grew up in Illinois and studied biology at the Illinois Wesleyan University graduating in 1912. He then spent two years as a laboratory instructor and postgraduate at the University of Louisville.

Becoming interested in pathology he studied that discipline in the Medical Faculty of Columbia University and was awarded his doctorate in 1916.

In World War I, Mueller volunteered to serve in France with a medical unit and surviving uninjured he was appointed an instructor in bacteriology again at Columbia University in 1919 where he concentrated on growing cultures of pathogenic bacteria.

Mueller determined that bacteria grow due to the presence of chemicals in animal or plant tissues and surmised that the body also requires at least some of these substances. he also predicted that some of these substances were unknown to science and medicine.

It is due to this concept that he isolated and characterised L – methionine which he determined to be essential for the growth of some strains of streptococci.

Moving to Harvard Medical School (1923), Mueller established that different strains of streptococci have different requirements in the amino acids needed for growth and replication.

This work expanded into researching the diphtheria pathogen and contributed to the development of vaccines for diphtheria and tetanus.

Joseph-Louis Proust

Angers – born Joseph-Louis Proust, always known as Luis, (September 26th 1754 – July 5th 1826 proved, in 1793, that the relative quantities of the elements in a compound are invariant and so essentially devised the concept of a chemical formula. This theory is known both as *'Proust's Law'* and *'The Law of Definite Proportions'* (1793) and laid the basis for analytical chemistry.

Originally apprenticed to his father as an apothecary, he became a professor of chemistry in the Spanish city of Vergara before moving to Paris in 1780. He became famous for taking part in one of the very balloon ascents. Proust moved around a lot, returning to Spain in 1786 to teach in Madrid and then, in 1792, moved to Segovia to teach at the Royal Artillery School. A demanding person, he was encouraged to seek opportunities elsewhere and he returned to Madrid in 1799 before becoming an apothecary in his home town of Angers.

Proust had a deep interest in nutritional chemistry and proved that the sugar extracted from grapes was identical to the sugar found in honey. His most important achievement in this field which is relevant to this volume is his isolation and characterisation of the amino acid, L – leucine.

Pierre Jean Robiquet

Rennes-born Pierre Jean Robiquet (13th January 1780 – 29th April 1840) was one of the founders of the studies of amino acid through his discovery and isolation of L – asparagine with Louis Nicolas Vauquelin (page 88) in 1806.

In 1816, working with together with Jean-Jacques Colin, he also a compound that they called *éther hydrochlorique* but is now known as 1,2-dichloroethane which they promoted as a medicine for people recovering from fevers.

His research was much wider than that and he identified the red dye, alizarin, and, the yellow red dye, purpurin in 1826. These two dyes were extracted from the roots of the madder plants which are widely distributed in the Mediterranean region as well as large parts of Asia, Africa and America.

He characterised codeine in 1832 whilst studying morphine and opium. He was also one of four chemists, all working independently, to isolate and characterise caffeine in 1820 / 1821.

In 1830, collaborating with Antoine Boutron-Charlard, Robiquet identified a new molecule which they termed *amygdalin** and which was the first glycoside*** to be isolated. They could not fully characterise the compounds and this was achieved two years later by Friedrich Wöhler and Justus von Liebig (page 98).

In the 19th century, amygdalin was one of the medicines used to treat cancer but it showed little efficacy and is actually toxic as it breaks down in the body with one of the products being hydrogen cyanide.

* *Amygdalin* is found widely in nature, particularly in the seeds of bitter almonds, apples, apricots, cherries, peachs and plums.

** A *glycoside* is a molecule in which a sugar is bound to another functional group via the, so called, glycosidic bond. Glycosides play a great number of roles and many plants store chemicals in the form of inactive glycosides which can be activated by enzyme hydrolysis.

Ernst Schulze

Born near Göttingen, Ernst Schulze (31st July 1840 – 15th June 1912) is renowned for his studies of cholesterol and phytosterols but most importantly for this isolation and characterisation of three amino acids: L – glutamine, L – phenylalanine and L – arginine. He also established the importance of both L – asparagine and L – glutamine in the proteins' metabolism in plants.

Schulze studied chemistry at the University of Göttingen under Friedrich Wöhler and then moved to the University of Heidelberg where he was taught by the world famous chemist, Robert Wilhelm Bunsen.

Schulze completed his postgraduate studies at the Friedrich Schiller University in Jena before moving to an agricultural research station near Göttingen. In 1872, he was appointed professor of chemistry at the Zürich Polytechnic where he remained for the rest of his career.

Towards the end of his career, Schulze studied the role of carbohydrates in the membranes of plant cells and can reasonably be described as one of the founders of biochemistry which, in his day, became known as *'physiological chemistry'*.

Adolph Strecker

The German chemist, Adolph Strecker (October 21st, 1822 – November 7th, 1871) is mainly remembered for his work in organic chemistry. Strecker was the first to synthesis L – alanine by the hydrogen cyanide catalysed reaction of ethanal with ammonia.

Born in Darmstadt, he studied science at the University of Giessen, where Justus von Liebig was a professor (page 98).

Whilst at Giessen, Strecker investigated both organic and inorganic chemistry especially the reactions of lactic acid, determined the atomic masses of both silver and carbon and determined how to separate cobalt and nickel which are often found together.

In 1851, Strecker became a professor of chemistry at the University of Christiania in Norway where he concentrated on natural products.

In 1860, Strecker moved to Tübingen University in Germany and then moved to University of Würzburg in 1870 where he continued his work on amino acids.

Louis Nicolas Vauquelin

Louis Nicolas Vauquelin (16th May 1763 – 14th November 1829) was both an apothecary and a chemist and is best known for his discoveries of chromium and beryllium.

Born in Normandy he became apprenticed to an apothecary in Rouen but later moved to Paris.

Another extremely prolific writer, he is credited with having written over 370 papers.

Surprisingly for an inorganic chemist, he was also interested in hens' eggs and plants. In 1806, whilst working with asparagus, he isolated and characterised the amino acid L – asparagine which was also the first amino acid to be discovered. He also discovered and isolated pectin and malic acid from apples.

Richard Martin Willstätter

Karlsruhe – born Richard Martin Willstätter (13th August 1872 – 3rd August 1942) and Nobel Laureate in Chemistry (1915) concentrated on the structure of plant pigments including chlorophyll and is credited with inventing paper chromatography, independently of Mikhail Tsvet who is more commonly credited with inventing the technique.

Willstätter studied at the University of Munich where he remained for fifteen years before moving to the Swiss Federal Institute of Technology in Zürich (ETH Zürich). Willstätter's doctoral studies focused on the structure of cocaine and other alkaloids. It was during this time that he also discovered and isolated, the amino acid, L – proline.

In Zürich, he concentrated on the structure of chlorophyll and was the first person to establish its empirical formula.

Moving to the University of Berlin in 1912, he continued his research on chlorophyll and is the first person to establish that it comprises two compounds, *chlorophyll a* and *chlorophyll b*.

In 1916 he returned to Munich where he studied the mechanisms of enzymes and established that enzymes are chemical substances and are not biological organisms.

Friedrich Wöhler

Born in Eschersheim, Friedrich Wöhler (1800 - 1882) was educated at the Frankfurt Gymnasium, Marburg University and Heidelberg University. At Heidelberg, he came to the attention of Leopold Gmelin who arranged for him to study under Jacob Berzelius in Stockholm. He served twenty one years as Professor of Chemistry at the University of Göttingen

Wöhler became famous for his discovery and isolation of beryllium and yttrium and his production of silane (SiH_4) and silicon nitride (Si_3N_4) as well as the first chemist to isolate aluminium in metallic form. He is, however, mainly renowned for working with Justus von Liebig, Pierre Jean Robiquet and Ernst Schulze and mostly for his laboratory synthesis of urea, $CO(NH_2)_2$.

Since urea is such a small molecule this now sounds trivial but the discovery was utterly seminal in organic chemistry. Until Wöhler synthesised the compound from inorganic compounds it was believed that some compounds, termed organic, could only be produced by living organisms and that these compounds has a so called, 'life force', or *vitality*. The latter term is important as it is the origin of the word *vitamin* which was originally a contraction of the term *vital amine*. Whilst we are all familiar with the word *vitamin* it was abandoned, technically, when it was found that not all vitamins are amines and they could all be produced in a laboratory from inorganic materials so were not *vital* either.

Together with von Liebig, Wöhler also had a considerable impact on the concepts of structural isomerism notably that compounds of the same molecular formula could have utterly different properties and behave completely differently. Wöhler and von Liebig exemplified this with silver fulminate and silver cyanate. Both compounds have the same elemental formula but whilst silver fulminate (AgCNO) is explosive, silver cyanate (AgNCO) is extremely stable.

Wöhler continued this work by demonstrating that heating ammonium cyanate, $[NH_4]^+[OCN]^-$, converts it into its isomer, urea, $CO(NH_2)_2$.

Wöhler's contributions to chemistry cannot be overestimated.

William Hyde Wollaston PRS FRS

William Hyde Wollaston PRS FRS (6th August 1766 – 22nd December 1828) is most famous for his discovery of palladium and rhodium and for determining how to process platinum in industrial quantities which made him very wealthy.

Born in Norfolk, as one of seventeen children, he studied science at Cambridge (Gonville and Caius College) and also obtained a doctorate in medicine.

More interested in chemistry, Wollaston stopped practising as a doctor in 1800 upon receipt of a substantial inheritance.

Elected a Fellow of the Royal Society in 1793 he became its president in 1820.

In 1810 he isolated and characterised L – cysteine, a non-essential amino acid and which is one of the very few sulfur-containing amino acids.

www.ingramcontent.com/pod-product-compliance
Lightning Source LLC
Chambersburg PA
CBHW081811220526

45467CB00006B/2160